Agrotechniques of Medicinal and Aromatic Plants

Agrotechniques of Medicinal and Aromatic Plants

Editors

Dr. V. K. Agarwal, Professor & Head
Dr. S.D. Upadhyaya, Professor
Deptt. of Plant Physiology
JNKVV, Jabalpur- 482004 (M.P.)

SSPH

SATISH SERIAL PUBLISHING HOUSE

403, Express Tower, Commercial Complex, Azadpur, Delhi - 110033 (INDIA)
Phone : 011-27672469, 27672852 Fax : 91-11-27672046
E-mail : ssp@ndf.vsnl.net.in, hkjain@satishserial.com
Website : www.satishserial.com

Published by :

SATISH SERIAL PUBLISHING HOUSE

403, Express Tower, Commercial Complex, Azadpur, Delhi-110033 (INDIA)
Phone : 011-27672852 Fax : 91-11-27672046
E-mail : info@satishserial.com, hkjain1975@yahoo.com

Print Again : 2023

ISBN : 81-89304-13-5

Composed, Designed & Printed in India

ABOUT THE EDITORS

The contributors to this book are leaders and experts in their fields and are actively engaged in the promotion of medicinal plant research and development.

Vinod Kumar Agrawal: Vinod Kumar Agrawal, Ph.D. (Plant Physiology) is a Professor and Head, Department of Crop & Herbal Physiology, Jawaharlal Nehru Krishi Vishwa Vidhyalaya (JNKVV) Jabalpur, Madhya Pradesh, India. Basically he is a Plant Physiologist, having research, teaching and extension experience of more than 35 years. He is a leader of Medicinal & Aromatic Plats R & D Unit at JNKVV, Jabalpur that has name and fame in collecting, conserving and domesticating valuable medicinal biodiversity not only in the State but also in the country. Recently ICAR has sanctioned a niche area of excellence in medicinal & Aromatic Plants to JNKVV under the leadership of Dr. Agrawal. About fifty research papers and many popular articles have been contributed by him in the field of Medicinal & Aromatic Plants. He has been instrumental in organizing Trainings, Workshops, Seminars and Symposia on Medicinal & Aromatic Plants as secretary; hence playing a key role in of Agricultural Human Resource Development (AHRD) to promote medicinal plant based crop diversification.

Sunil Dutta Upadhyaya: Ph.D. (Botany) with specialization in Ecology is a Professor (Agril. Botany & Crop Physiology) Deptt. of Crop & Herbal Physiology, JNKVV, Jabalpur, MP, INDIA. He did his Postdoctoral degree from University of Florida, USA in Forestry (1987-1988) and Advance training in Agroforestry Management from University of New England, Australia (1997). His significant contribution in wasteland development through agroforestry at Madhya Pradesh include his direction as PI of the project on Technology Development, Extension in Agro forestry for Wasteland Development sponsored by Ministry of Rural Development, GOI, New Delhi. He has 25 years of experience in teaching, research and extension in the field of Medicinal Plants & Agroforestry. Presently he is PI of ICAR Network Project on "Impact, Adaptation and Vulnerability of Climate Change on Agro-biodiversity ". He has 40 research papers in his credit in referred Journals of Plant Science.

PREFACE

India has the treasure of medicinal plants due to wide variability in agro climatic conditions spread all over the country. Medicinal plants occupy special position in spiritual, social and cultural lives. We firmly believe in the inherent properties of medicinal plants to ward off diseases and to promote health.

India has 16 agro-climatic zones having 45000 different plant species in which 15000 species are medicinal that include 7000 plants used in Aurveda, 700 in Unani, 600 in Sidha medicines, 450 in Homeopathy and 30 in modern medicine. With this heritage of medicinal plants one-day India will become one of the richest country in the world regarding herbal wealth.

The WHO has estimated that 80% of the populations of developing countries rely on traditional medicine, mostly derived from plants for their primary health care needs. The demand of medicinal plants is increasing throughout the world. We have realized the adverse effects of synthetic drugs and there is perceptible change towards the liking for herbal medicines; because :

(1) Herbal medicines stood the test of time for their safety, efficacy, cultural acceptability and lesser side effects.
(2) The chemical constituent present in them are parts of physiological function of living flora and hence they have better compatibility with human body.
(3) Availability of sufficient literature mentioning the effectiveness of herbal medicines for age related disorders namely memory loss, osteoporosis, diabetes and liver disorder etc. for which only palative therapy is available in modern medicines.

Ninty percent of the drugs used in Indian system of medicines are plant based and collected from wild sources. This natural resource base is speedily shrinking day by day due to the reasons known to all. Therefore there is a need for conservation and sustainable use of medicinal plants. Cultivation is clearly a sustainable alternative for the present collection of medicinal plants from wild resources. Users, farmers, growers, entrepreneurs, extension worker and researcher require accurate contemporary and correct information for further horizontal expansion of the knowledge.

Considering the importance of medicinal and aromatic plant species, the present book on ***Agrotechniques on Medicinal and Aromatic Plants*** is mainly designed on cultivation practices, post harvest practices, quality evaluation and marketing aspect of medicinal and aromatic plants.

We hope this book will provide the necessary technical know how to motivate the interested growers and farmers to take up cultivation of medicinal plants.

We express sincere gratitude to the contributors and to our honourable Vice Chancellor, Director Extension Services, Director Research Services and Dean for their inspiration, guidance and keen interest and special thanks to staff members of the Department of Plant Physiology for their active cooperation in preparation of manuscript.

V.K. Agrawal
S.D. Upadhyay

जवाहरलाल नेहरु कृषि वि□वविद्यालय
कृषिनगर, अधारताल, जबलपुर - ४८२ ००४ (म.प्र.), भारत
Jawaharlal Nehru krishi Vishwa Vidyalaya
Krishinagar, Adhartal, Jabalpur - 4182 004 (M.P.), NDIA

डॉ. ध्यान पाल सिंह
कुलपति
Dr. D. P. Singh
Vice-Chancellor

Ph. : 0761-2481706 (O) 2480789 (R)
Fax : 0761-2481389, 2481074
E-mail : dpsingh_jnkvv@yahoo.co.in

FOREWORD

The Indian subcontinent represents one of the greatest emporia of medicinal and aromatic plant wealth with a vast scope and unparalleled opportunity for profound research and extension in the field of cultivation, processing, value addition and marketing of these plants. It has also been estimated that more than 80% of human population of developing nations rely upon medicines derived from plants resulting into heavy demand of medicinal plant materials at global level. Medicinal plants in India has considerable reputation as drugs; and it is pertinent to identify, domestication and multiply these valuable plants before they extinct from natural habitat. But these natural resources of medicinal value are shrinking rapidly day by day, which causes concern. Fortunately India has diverse climatic conditions and physiography resulting into rich flora of these plants in diversified forest area. The State of Madhya Pradesh has also blessed with rich floral wealth due to its broad geographical identities, climatic variability together with diverse soil types; providing unique opportunity to grow economically viable medicinal and aromatic plants in marginal and culturable waste lands. Now, the time has come to conserve these plant wealth through in situ and ex situ conservation for its full potential benefits to the mankind.

I am happy that this book, **Agro-techniques of Medicinal and Aromatic Plants** is a well-designed compendium of research findings, considering the global importance of these endangered species of medicinal value. The good efforts of the scientists of the Department of Plant Physiology of this Institution and the Editors and Members of Editorial Board to bring out such an important publication is highly appreciated. I hope, this compendium will be useful to the planners and field workers those who are engaged in boosting up the cultivation of these important crops not only at State level, but also at National level.

(D.P. Singh)

CONTENTS

AONLA
(*Emblica officinalis* Gaerth.)

A.B. TIWARI

Scientist

Deptt. of Plant Physiology, JNKVV, Jabalpur (M.P.)

Botanical name	:	*Emblica officinalis* Gaerth.
Family	:	Euphorbiaceae
Common name	:	Amlaki, Anola, Amlika, Aonla, Amrit Phal

Introduction

Aonla (*Emblica officinalis* Gaertn) belongs to the family Euphorbiaceae has been recognized as 'Amrit Phal'. It is probably the only fruit to fill the gap of astringent food recommended by Aurvedic System of medicine for balance diet and sound health. It has high value among the indigenous medicines in India. The fruit contains a chemical substance known as gallic acid (polyphenols), which retards the oxidation of vitamin C, hence it has antioxidant property. Fruit is therefore rich source of vitamin 'C' in both fresh as well as in dry or processed form. Fruit is acrid, cooling laxative and diuretic.

Climate and Soil

The plant occurs throughout tropical and subtropical India. The tree is very productive, hardy and can be successfully grown in variable agro climatic and soil conditions. The mature tree can tolerate temperature as high as 46^{0}C as well as freezing temperature. Thus it is not much influenced by hot winds and various kinds of degraded lands such as salt affected soils, ravines and dry and semidry regions.

Varieties

Commercial varieties of aonla such as Banarasi, Francis and Chakauya have some limitations. Banarasi is a shy beared, Francis is affected by necrosis and Chakaiya bears small sized fibrous fruit. These are the new varieties developed by NDUAT from the selection.

Narendra -6 This is a selection from Chakaiya. It has mid season maturity. Fruits are most attractive and shining. Medium to large sized, flattened and very low in fibre content. Heavy bearer and most suitable for candy, preserve, sweets, Jam and sauce making.

Narendra - 7 This a selection from Francis. It has also mid season maturity. Fuits are medium to large sized with conical apex and free from necrosis. Fibre content is little higher than NA-6. This is most precocious and prolific bearer. Very good for Chaywanprash, Chutney, pickle, Jam and Squash making.

Narendra -10 This is a selection from Banarasi. It has an early marutiry. Fruits are medium to large sized, flattened with roundish stylar end. Fruits during early stage of development have pink tinged colour. Fibre content is similar to NA-7. This is also a heavy bearer and most suitable for dehydration and pickle.

Other improved varieties

Krishna (NH-4), Kanchan (NH-5).

Self-incompatibility has been observed in aonla, hence two varieties such as Narendra-6 + Narendra-7 or Narendra -7 + Narendra -10 or Chakaiya + Narendra - 7 be planted in alternate rows for higher productivity.

Propagation

Aonla is usually propagated by seed but seeding plants are not true to type and take more time to bear fruit. This has necessitated standardization of vegetative technique for commercial raising of plants. Budding has been found most successful technique. For this purpose it is essential to raise seedlings to be used as rootstock. Fruits of seedling aonla tree preferably Chakaiya or kanchan (a selection from Chakaiya) should be used for extraction of seed. One quintal of fruits can give about one kg. seeds.

Seeds should be soaked for about 12 hours in water before sowing. Seeds should be sown on raised beds during March-April. Early sowing in February can also be done in poly house. Seedlings of late sown seed (May-June) do not reach to buddable size in same season; hence they are used for next season. Seeds sown during March/April germinate within two weeks and seedlings of about 10 cm height (after 35-40 days of sowing) after holding 3 days in shade should be planted for better establishment. Seedlings of about pencil thickness (90 days old) can be used for budding. The scion shoot should be selected from mother plant, which are prolific bearer. Patch budding proved to be the best method, which gives highest success (90-100 per cent) when performed in June-July. Plants are ready for sale or planting in August-September. The cost for raising budded plant is about Rs. 4/plant.

Planting

Budded plants should be planted at a distance of 8x8 to 4x4 metre in square system depending on soil type. Pits are dug in normal size 1 x 1 x 1 metre. Care should be taken to remove the hard kankar pan of salt affected soils. Pits should be filled with the mixture of normal soil + 25kg well rotten Farmyard manure + 100g BHC upto a height of 6 to 10 cm above the ground level. Irrigate the pits to settle down the soil.

Planting can be done in late January or February in spring and July to September in rainy season. Plant survival and growth have been found better in spring season planting (under dormant condition) with assured supply of irrigation. The plants with its earth ball should be taken and be placed in the center of the pit by excavating as much soil as to accommodate the earth ball. It is better to adjust at same depth at which it was in nursery. The moist soil of the pit is then pressed around the earth ball. Light irrigation should be given just after planting.

Pruning

Young plants are allowed to grow straight in order to provide good framework of the individual tree. The framework should be developed by encouraging the growth of 4-6 well-spaced branches on the trunk at a height of 60-75 cm, from the ground level. Pruning in aonla is normally not practiced. Dead diseased, weak, crossing branches and suckers appearing from rootstock are pruned off.

Manures and fertilizers

Little or no systematic work has been done on nutritional requirement of aonla fruit. Plant produces heavy crops hence; application of manure and fertilizer will be beneficial. Recommendations based on experience gained are as under-

Apply 10 kg of farmyard manure, 100 g nitrogen 50 g phosphorus and 75 g potash per plant to one-year-old plantation. This dose should be increased every year in same proportion up to the year of 10 years. Farmyard manure should be applied in the month of January. Half dose of nitrogen and potash and full dose of phosphorus should be applied during February i.e. before flowering. Remaining half dose of nitrogen and potash should be given during the end of August. The manure and mixture of fertilizer should be spread under the whole canopy of the plant and it should be incorporated well in surface soil by spading. Irrigation should be given immediately.

Irrigation

Plant need to be cared for watering. Basin system, which provides more uniform distribution of water, should be used for irrigation of young plantation. This system of irrigation also minimizes the movement of salt into basin hence it has special advantage for those plantation, which has been done, in salt affected lands. Young plantation should be irrigated at 10-15 days interval during summer. In bearing plantation first irrigation should be given just after maturing and fertilizer application. Irrigation should be avoided during flowering. Irrigation at 15-20 days interval should be given after fruit set (April to June). Aonla Plantation is highly susceptible to water logging, hence care should be taken to save the plantation from excess or water logging condition during rainy season.

Insect pests

The major insect pests of aonla are bark eating cater pillar (*Inderbel tetraonis*), shoot gall maker (*Betousa stylophora*), scale insect and anar butterfly (*Virachola isochola* isocrates).

Bark eating caterpiller affects the trunk and branches seriously. Injecting kerosene oil or endosulphon (0.05%) in holes and plugging the hole with mud has been found effective to control this insect.

Shoot gall maker affect the apical shoot of young plant and trees. This can be controlled by pruning the galled shoot and spraying of monocrotophos (0.05 %) during rainy season.

Scale insect damages the growing shoots and leaves by sucking the sap. This can be controlled by spraying of 0.05 % monocrotophos.

Anar butterfly causes considerable damage to the fruits. Remove the affected fruits and destroy it. Spray monocrotophos (0.05%) at the time of fruit development to control this pest.

Disease

Among the diseases, Aonla rust caused by a fungus (*Revenellia emblicae*) is of the major importance. Brown pustules appear on both leaves and fruits and finally these become dark brown to black. Affected fruits drop before maturity. Spray of Diathane M-45 /Indofil M-45 (0.35 or 3g/lit) twice has been found effective to minimize the problem. First apray should be given in beginning of September and second spray after 15 days. About 10 litres solution will be required for a full-grown tree.

Harvesting

The stage of maturity at which fruits are harvested has greatest effect on quality, storage life and marketability. Early and late harvesting both causes poor shelf life. Therefore, fruit should be harvested at full maturity (full growth) which varies from variety to variety. Early variety such as NA-10 and Banarsi should be harvested in November. Mid season variety such as NA-6 and NA-7 should be harvested after middle of December and late variety such as Chakaiya in January in eastern part of Uttar Pradesh.

Yield

Bearing starts in budded plants after 4-5 years of planting and tree starts to give full commercial yield at the age of 10 years. Average yield at this age is about 80-100 kg per tree or 11 to 14 tonnes/ha.

Intercropping

The long gap between planting of orchard tree and its production is the biggest problem for the poor farmers to adopt plantation on large scales. Therefore, inter spaces between the plants rows must be utilized for additional crops. This may not only provide some additional income to the growers in early years but also provide maximum soil cover. Among the fruit crops guava and Karonda can be used successfully as filler in aonla plntation. Vegetable crops such as bottle gourd, bhindi, cauliflower, coriander etc, medicinal crop such as kalmegh, ashwaganda, chandrasoor, senna, satawar, safed musali, germen chamali etc flower crop such as gladiolus or marigold can be grown in association with aonla plantation.

Chemical analysis

Aonla fruit is one of the richest sources of vitamin "C" also light content of tannin i.e. gallotanic acid, carbic acid, glucose, pectin, Root content Ellagesic acid, Loopiol. Seed content 16% of oil. Leaves flower and bark also content tannin.

Fruit Analysis

Vitamin C	-	600 to 921-mg/100 Gms pulp
Water	-	82%
Proteins	-	0.5%
Fats	-	0.1%
Salt	-	0.7%
Fibers	-	3.0%
Sugar carbohydrates	-	15 to 20%

Others constituents are Phosphorus, Iron, Fe, Nicotinic acid, Carrotin-C, Thymin, Rhiboflavin, Nigacine, Mitheonin, Traptophen.

Uses

Fruits are aperient, aphrodisiac, astringent digestive, diuretic, laxative, refrigerants and tonic. Useful in anaemia, jaundice, dyspepsia, haemorrhagic, disorders, bilionsness, asthma, bronchitis, diabetes, urination, certain heart complaints, diarrhoea and dysentery.

Flowers, root and bark of the tree are also possesses medicinal value. Seeds cure asthma and stomach disorders.

Aonla fruit has great potentiality for processing into number of quality products, such as Pulp, Jam, Herbal Jam, Jelly, Toffee, Sauce, Herbal syrup, Candy shreds, Pickle, Aonla sweet, Aonla powder, Diabetes powder, Triphala, Chyavanprash etc.

Fruits are largely used in inks, dyes and for hair shampoos.

Leaves are used as manure in *cardamom* plantation; they help in improvement of alkali soils.

Timber is useful for miscellaneous domestic purposes. It stands well under water.

BIJASAL
(*Pterocarpus marsupium* Roxb.)

S.D. UPADHYAYA
Professor, **Deptt. of Plant Physiology, JNKVV, Jabalpur (M.P.)**

Botanical name	:	*Pterocarpus marsupium* Roxb.
Family	:	Papilionaceae
Vernacular name	:	Hindi- Bijasal, Sanskrit – Mahakutaj Gujarat- Biyo, Hiradakhan Tamil- Vaengai

Morphological characteristics

It is a large tree. Leaves compound, having 5-7 leaflets (8-13 cm), long, oblong or elliptic. Flowers about 1-5 cm long, yellow, scented in very large, dense branches. Fruit 2-5 cm long, roundish, winged, with one to two seeds.

Habitat

The tree is common in central and peninsular India, chiefly found in mixed deciduous forests of Gujarat and Madhya Pradesh.

Geo-climatic conditions

It prefers fertile, deep soil condition for its growth and development. It thrives well under clear-cut differentiated summer and winter with medium rainfall during rainy season (700-1250 mm). It is light demander, tolerant of excessive temperature in summer.

Soil and land preparation

Deep clayey loam to vertic soil (black soil) with good drainage is best suited for growth and development of *Pterocarpus marsupium*.

Recommended varieties

The varietal improvement work is not reported in India. However, provenances collected from Amarkantaka had shown better growth.

Planting material

Propagation through seed is good, but seed viability and longevity is very less. Attempts are made to develop micro-propagation technique of Beejasar through tissue culture of axillary buds and leaf disc. Callus formation is very good, but rooting is not induced yet. Freshly collected seeds should have given water soaking for 48 hrs before sowing gave good germination.

Method and time of raising the plants

Best treatment recommended: Preparation through axillary buds and leaf discs using micropagation/tissue culture techniques.

Optimum spacing recommended: 8m x 8m for intercropping.

Best period for sowing/planting germplasm: 15th June to 15th July.

Viability of germplasm

Seed viability - 3 months after maturity. It is very prone to seed borer attack which causes serious loss in seed viability. There is a hidden syndrome of dormancy may be due to hard seed coat and presence of germination inhibitors is seed coat.

Recommended seasonal practices and intercultural operations:

Procedure		Periodicity
a) Pit digging	-	April - May
b) Transplanting	-	15 June - 15 July
c) Weeding	-	Two weedings: First one and second four months after Planting
d) Irrigation	-	Every two month in winter and 30 days interval during summer.
e) Gap filling	-	In September

Maximum number of plants appropriate for plantation per hectare through cropping / inter cropping system

156 plants per ha with provision of intercropping between rows.

Manure and nutrient requirements including time and method of application

25 kg well decomposed FYM 200 gm nitrogen and 150 gm phosphorus per plant per year for three years. The nitrogen should be applied in two splits first in September and second in January.

Irrigation practices including quantity and frequency of application

6 irrigations are recommended (November, January, March, April, May, June) through check basin methods.

Weed control method including chemical control

Two weeding one in August and second in November through hand weeding. The hand hoe can be used for weeding in leveled land while in undulating terrain sickles can be used for hand weeding.

Diseases, insect pests, nematodes, physiological disorders, if any and their control measures

There is no serious insect pest and disease attack seen in mature stem and roots. However, seeds are very prone to seed borer insect, which is main reason behind early loss of seed viability. This may be controlled by proper drying (upto 12% moisture) and use of carbon disulphide in storage. In nursery, and early plant stage, leaf eating and white grub attack is reported, which can be controlled by floor sprays of Endosulphan @ 0.03% and application phorate 10 G near root respectively. In order to keep plant disease free in nursery and early stage, the seed treatment, using thiram @ 3 gm/kg seed, is must.

Time of flowering and fruiting

Flowering	-	November-December
Fruiting	-	February-March

Period required for maturity

3 months for fruit maturity and 8 years for plant maturity.

Harvesting

October month (6-8 year after transplanting) i.e. just before flowering, because it is well established fact that plant have maximum active principle content in their specific part, which can be reallocated or partitioned towards flowers and fruits later on. Fruits mature in 3-4 months, while plant mature for harvesting in 6-8 years growth after development of heart wood in stem, i.e. medicinally important plant part.

Parts used for medicinal purpose

Leaves, flower, gum and heartwood are used.

Kino from incisions in bark - astringent, antidiarrhoeal, anti-haemorrhagic, in pyrosis, haematuria, tongue diseases, locally in leucorrhoea.

Heart wood: Antibiotic, hypoglycaemic and used to control blood sugar.

Flower and Kino from bark: Febrifuge.

Leaf: Used externally on boils, sores and other skin diseases.

Concentration of active constituents in different parts of the plant

Isoflavonoids, terpenoids and tannins are reported from heartwood and roots of *Pterocarpus marsupium*. Heartwood gave (-)-epicatechin, 4,4'-dihydroxy-amethylhydrobenoin (marsupol), 5,4'-dimethoxy-8-methyl-isoflavone, retusin-7-glucoside, irisolidone-7-rhamnoside, 5,7-dihydroxy-6-methoxy-isoflavone-7-rhamnoside, carpusin, proterols-A and B. Roots gave liqui-ritigenin, garbanzol, isoliquiritigenin, garbanzol, isoliquiritigenin, p-hydroxybenzaldehyde, pterosupin, pseu-dobaptigenin and 5-deoxy-kaempferol.

Other relevant information

It is very difficult to propagate through seed, because of rapid loss in seed viability and strong dormancy constraints. Micro-propagation using vegetative plant part is also difficult due to no success in rooting induction after callus formation through tissue culture method.

SINDURI (*Bixa orellana* Linn.)

SHARAD NEMA
Deptt. of Plant Physiology, JNKVV, Jabalpur (M.P.)

Botanical Name : *Bixa Orellana*
Family : Bixacea
Common name : Sinduri (Sanskrit), Sinduriya Latkan (Hindi), Arnotta plant, Annatto plant

Distribution

The plant is native of tropical America and the West Indies. In India, the annatto plant is found in Karnataka, Tamil Nadu, Andhra Pradesh, Orissa, Maharashtra, West Bengal and Assam. However, India exports annatto seeds to Russia. It is grown as a hedge-plant or a roadside tree in many places.

Identification of the Plant

Annatto is a shrub or small tree which grows to about 5 m high; the leaves are cordate, stalked, acuminate, sometimes with reddish veins; the flowers are occur in terminal, erect panicles; the sepals are 5, deciduous; and there are 5 petals.

This species occurs in two forms, one with white flowers and green capsules and the other with pink flowers and red capsules. The capsule is heart-shaped, covered with soft spines, the seeds are numerous, about 5 mm in diameter with a thin, pulpy, vermilion skin.

Soil

It is a hardy tree and does not require very fertile soils for its cultivation. An average soil, which has some moisture-retaining capacity, is suitable for the cultivation of this crop. Any soil, which has poor drainage and rocks in the subsoil, should be avoided.

Climate

Annatto is a tropical plant. It can be grown successfully in a warm, dry climate. It cannot withstand frost.

Season

The best time for raising the nursery is during the months of April and May. One-year-old, healthy seedlings can be planted in the main field at the onset of the rains.

Land Preparation

The land is levelled and prepared well before planting. Later, pits of 30 cm x 30 cm x.30 cm size are dug in early March at a spacing of 4.5 m x 4.5 m. Th-e pits are filled before the onset of the monsoon with a mixture of soil, compost and fertilizers.

Cultivation

Propagation

The plant can be propagated by seeds or vegetatively through mature stem-cuttings. Generally, the seedlings are raised in polythene bags containing a mixture of soil, manure and fertilizers. The seeds germinate in 8-10 days and only one seedling is grown in each bag. The seedlings can be transplanted when they are about 20 cm tall.

Planting

The seedlings are planted at the centre of the pit and the pits are watered immediately after planting. At the spacing recommended above, about 500 seedlings can be planted in a 1 ha area.

After-care

The weeds are removed around the plants in the pits as and when they appear. Light digging in the pits is also done after every weeding to encourage good growth of plants.

The plants start flowering in the first to second year after planting. To encourage good vegetative growth and for better performance of the plantation in subsequent years, the flowers are plucked off in the first two years.

Pests and Diseases

The plant is very hardy in nature and is not attacked by any pests or diseases of a serious nature.

Flowering, fruiting and harvesting

The plants start bearing fruits after 3 years of planting and continue to give economic yields up to the age of 8 to 10 years under good management. The flowering season and also the habits are reported to differ from the hills to the plains and the forms of the species. Generally, the plants flower in the last week of August and the flowering continues up to the middle of October. In about 30 days, the flowering capsules are borne on the trees and after 90 days appear in the form of capsules. The capsules will mature and dry up in January. The dried capsules make a rattling sound. Another indication for harvesting the capsules is the formation of a crack at the stigmatic end of the capsule. The capsules should be harvested at this stage. If the capsules are allowed to burst, there is a heavy loss of seeds. After harvesting the panicles, they are packed in gunny bags, which are kept closed for some days. Later, they are taken out and exposed to the sun. The seeds are separated from the dried capsules by beating them with sticks. The seeds are sun-dried, cleaned and packed in gunny bags before they are marketed.

Yield

A 3-year-old plant will yield, on an average 1/2 kg of seeds and 1 kg /tree/year thereafter up to 10 years of age. The seed-yield from a 1 ha plantation will be around 250 kg from a 3-year-old and 500 kg from a 4- to 10-year-old plantation.

Extraction of dye (Annatto colour)

Several methods have been described and patented for the extraction of crude annatto colour and bixin from the seed. One such process of extraction is by the counter-current method, using water during simultaneous mechanical friction of the seed surface. Rnally, the suspension is evaporated in a vacuum and subsequently dried and ground to a fine powder.

The other method of extraction involves the treatment of seeds with a vegetable oil, with or without heating the mixture and also treating the seeds after swelling them in water or in steam. Another method of extraction is by boiling the seeds with sodium carbonate solution, filtering _md acidifying the filterate. Later, the dye is coagulated by boiling it with common salt, after which it is filter-pressed, washed and dried. In India, a better colour is manufactured by heating the seeds under pressure and then extracting the dye with a vegetable oil, preferably castor oil. It has been observed that the pre-treatment of seeds is not required for extraction by this method.

Bixin can be separated from the seed by boiling the seed with ethyl acetate or acetone or alcohol or chloroform. Bixin can be held in alcohol solution at a pH of 4.5 for one hour before spray drying. The commercial cheese-colour is an alkaline solution and contains norbixin in place of bixin. It is extracted by using dilute sodium or potassium hydrochloride.

The stabilization of the annatto food colour is done by the addition of curcumin, vanillin, eugenol or vitamin E.

Uses

Annatto is a natural carotenoid obtained from *Bixa orellana* Linn., a shrub. In India it is called Rangmale, Latkan, Japhra or Sandri. The seeds of this plant are gaining economic importance in recent years due to ediable and desirable colourant. The ripe fruits of annatto, on drying, yield annatto seeds, which serve as a raw material for the production of annatto colour. The seed coat consists mostly of carotenoid dyes (about 2%) and an essential oil. Bixin (10-30%), which is the principal colouring matter present in seeds along with small amounts of 'Orellin', a water soluble yellow substance, fat, resin and a bitter stuff. The annatto dye is non-toxic and is mainly used for colouring edible materials like butter, ghee, other milk products, chocolate and cosmetics. It neither affects the colour nor the aroma coloured coating materials have been made using annatto and cellulose acetate phthalate and the same is used in coating tablets, pills and granules. The addition of annatto seed-meal to poultry feed is reported to deepen the colour of the egg-yolk and also the flesh pigmentation. Annatto dye is also used in floor polishes, shoe polishes and hair oils.

NILGIRI (*Eucalyptus* spp.)

AASHUTOSH SHARMA
Research Associate
Deptt. of Plant Physiology, JNKVV, Jabalpur

Botanical Name : *Eucalyptus*
Family : Myrtaceae
Common Name : Nilgiri, Safeda

Introduction

Eucalyptus spp. is native of Australia and they are introduced in India. The precise date of introduction of *Eucalyptus* in India is not known but according to Troup a few trees most probably *Eucalyptus globulus* were planted in the Nilgiri Hills in 1843. Small plantations of *E. hemiphloia, E. pyconantha, E. eugenioides, E. pilularis, E. punctat, E. propinqua, E. acmenoides, E. paniculata* and *E. crebra* in existence in the Nilgiri hills. Since then, a few other species of *Eucalyptus* have been tried in different states of India primarily with a view to provide firewood and raw material for paper pulp industry on a relatively short rotation. *Eucalyptus* has shown wide amptitude of adaptability. From a review of literature, the main species adapted to the different regions of the country are *E. grandis* in the southwest, *E. tereticornis* in the center and *E. globules* at the high elevations where there is no snow. On swampy and slightly saline soils, *E. robusta* is the most suitable species.

Eucalyptus globules popularly called "blue gum" is extensively cultivated for the extraction of oil from the leaves and as a source of pulp wood and fuel.

Eucalyptus citriodora Hook, has been placed as the second most important species of *Eucalyptus* for essential oil trade only after *E. smithii*. *E. citriodora* Hook (syn *maculata* Hook, variety citridora Bailey) is a large tree often attaining a great height with a smooth, whitish to pale pink bark. It is commonly known as citron scented gum or lemon-scented gum. It can be easily identified by its characteristic fruit and lemon scented leaves. Cultivation of *E. citriodora* is rapidly being extended throughout the world. Though a little slower in growth than others and somewhat frost sensitive in seedling stage is widely planted throughout India and thrives well even on quite difficult sites. In recent years extensive plantation of *E. citriodora* has been undertaken in India, noteworthy centers are Punjab, Uttar Pradesh and Andhra Pradesh.

Climate

This plant is sensitive to severe drought and frost. However, it can tolerate arid conditions very well. It can be grown under rainfall ranges from 30 to 40 cm per annum up to the 4000 mm per annum. The tree does well in the elevation up to 1500 to 2000 m MSL and thrives best in areas with considerable soil moisture but grows fairly well on dry ground also.

Soil

It can be grown in acidic soil, rich in organic matter and well-drained sandy-to-sandy loam soils. This can be planted in hilly shallow soils under fallow/waste lands.

Propagation

Propagation is by seeds. *Eucalyptus citriodora* can be grown as direct seeding as it has greater mortality during transplanting. However, careful transplanting of plants raised in poly bags can be survived. The sowing period of seed is January/ February. The seeds germinate within 10-15 days. The seedlings would be ready for field planting by May.

Planting

The planting should be done after 2 to 2.5 months during June-July under rainfed conditions. However, the transplanting under irrigated conditions can be done any time. The planting distance may be decided depending on the sole crop or as inter crop in the Agroforestry system and the purpose of the crop grown.

For oil extraction purpose the planting is done at distance of 1mx1 m plant to plant and row-to-row distance. For intercropping, it should be planted at a distance of 2.5mx2m and for sole crop for pulp and timber at a distance of 2mx2m. Staking of them is required during first year and gap filling is done till the second year.

Intercropping

Eucalyptus can be grown along with coffee, lemongrass and palmarosa. In the first four years, intercropping with pineapple, yam and vegetables can be done.

Manure and fertilizers

10 kg of FYM/plant during planting is applied. A fertilizer mixture of 25:38:38 g of NPK (200 g mixture) per plant is recommended by Cinchona Department of Tamil Nadu. It should be applied at the end of monsoon season.

Weeding

The weeds should be removed ones or twice during first two years of transplanting for optimum growth of the plants.

Irrigation

The plants should be irrigated after transplanting if rains not occur. The plants can be grown as rainfed. However, irrigation should be provided during first year in summer.

Harvesting

When plants are allowed to grow in tree form, a poor yield of foliage is obtained and at best it takes two years before the first crop can be taken. However, if close planting is done in spring and plants coppiced next summer, just before the onset of monsoon, formation of lateral shoots is promoted resulting in bushy growth and excessive leaf formation. Thus, three crops of leaf can be harvested every year. Based on seasonal oil yield, citronellal content studies, optimum leaf maturity considerations and cropping sequence, it is suggested that in N.W. India two major crops may be harvested in April and November while a minor third crop may be obtained in July. The harvested leaves are dried in the shade for one day and distilled. Another method is pollarding the main stem at a 3m height and the leaves are regularly collected from the shoots, which emerge from the pruned stem. The leaves yield 1% oil on efficient distillation.

In case of *E. globulus* leaves are harvested by cutting the side branches twice or thrice in a year. The fallen leaves are also collected from the plantation. Finally, the leaves are collected when the trees felled for pulp wood purposes. These leaves can be distilled throughout the year but most suitable period is April and September because the oil yield and Cineole content during this period will be more. The collected leaves are dried for 3 days then steam distilled. The distillation requires 5 to 7 hours per charge the yield of oil varies from 0.75 to 1.25%. The wet crude oil should be rectified by treating the oil over anhydrous sodium sulphate and distilling over 1 to 2% caustic soda. Cineole contents varies from 75 to 85%.

Chemical constituent

Leaf essential oil contains mainly camphene, sabinene, myrcene, P-menthane, a and r-terpinene, fenchone, a and B-thujone, Citral, Verbenone, Cineole, isoamyl alcohol, trans-pinocarveol, trans-verbenol, borbneol myrtenol, terpineol-4, a terpineol, eudesmol, thymol, aromadendrine etc.

Uses

The oil, which is obtained by steam distillation of the leaves, is widely used in soap, perfumery and cosmetic industries and for the isolation of hydroxy citronellal used in the manufacture of high-grade perfumes. Essential oil from leaves of *E. globulus* anti-asthmatic, antiseptic, antirheumatism, anti dropsical, antispasmodic, disinfectant, mosquito repellent, febrifuge, carminative, diaphoretic, rubefacient, stimulant, expectorant, in chronic bronchitis, migraine, and other farms of neuralgia, diptheria, ulceration of throat, skin diseases, leaf is vermifage and root is purgative.

ARJUN, KOHA (*Terminalia arjuna*)

AJAY KHARE
Assistant Professor,
Deptt. of Plant Physiology, JNKVV, Jabalpur (M.P.)

Botanical Name	:	*Terminalia arjuna*
Family	:	Combretaual
Common Name	:	Arjun, Koha

Distribution

Throughout the greater part of India, mostly along rivers, streams, running and dry water courses. Quite common in Chhota Nagpur, Madhya Pradesh, part of Maharashta Tamilnadu, Uttar Pradesh, Andhra Pradesh, Assam, West Bengal, Karnataka, Orrissa and Gujrat.

Climate

Max. Temp.	:	38^0 to 48^0 C
Min. temp.	:	1^0 to 15^0 C

Annual rainfall - 750 to 1750 mm. (Even 500 mm.). Requires moist cool places for its better growth and development. Does not thrive well in drought condition.

Soil

Grows on a variety of soils including saline and alkaline provided, good moisture supply. The best growth is found on neutral fertile soils specially loose moist alluvial loam with good water supply and drainage.

General Description

Almost evergreen, large sized (24 m height, 2 m dbh) tree with a broad crown and shallow roots. Clear bole is usually half of the total height, buttressed at the base, bark is gray or pinkish-green, smooth. Leaves are simple large, opposite, new leaves appear in April before all old leaves shed. Flowers are small, white to creamy white in spikes, appear in April to May. Fruits 2.5 to 6 cm. long woody, with 5-7 raised ribs ripening in the following February to May, dark brown to reddish-brown in colour moderate shade bearer, young plants are succeptible to drought and frost. It is fire tender.

Propagation

It can be propagated by seeds through direct sowing, seedling plantation and by air layering. Trees start fruiting after 6-7 year; every third year is a good seed year.

Seed collection

Ripe fruits are collected from the trees in April-May, seeds are encased in the bony fruits, about 175-450 fruits weight 1 Kg. Fruits are dried in the shade for a month or two and ribs are generally reduced by rubbing with the hands. The fruits store well for one year when stored in sealed tins. If stored in Gunny bags germination is reduced by 10-20%.

Seed treatment

The fruits are soaked in water for about 48 hours and then put in shallow pits or trenches till sprouting commences. Putting the fruits in boiled water allowed cooling and keeping them soaked in the same water for 72-96 hours improves and hastens germination. The use of pre-germinated seed is preferred to ensure uniform germination and to save time taken to germination.

Nursery raising

In nursery beds sowing is done in lines 30 cm. apart with a spacing of about 5 cm. between the seeds. A germination percentage of about 50 to 65 is reported from untreated seeds, while treated seeds may give 90%

germination. Germination commences in about 10 days and prolongs up to 2 months. 5 Kg. fruits yield 1000 plants. Regular weeding and watering is done as and when necessary.

The seeds are also sown in polypots in Feb.-March (two seeds per pot) at a depth of 2.5 cm. after germination; the plants are thinned to one seedling per polypot. Seedlings grow to 40 cm. height by July and becomes ready for planting out.

Direct sowing

Direct seeding is not recommended in dry areas, watering can be done to encourage germination, in case of a brink in the monsoon.

Stumps

Stumps are prepared from 15 months old seedlings. (root - 20 cm. and shoot 5 cm.).

Planting

Seedling is planted in July-August in pits of 30 X 30 X 30 cm size, dug in advance. Stumps are planted in pits or in crowbar holes. It has been successfully grown in taungya plantations and is suitable for canal and road side plantation, (5 to 8 m.). It is now largely planted for raising tasar silk cocoons, spacing adopted is 1 X 2 m. and management by repeated pollarding to provide tender leaves to tasar worms.

It is initially slow growing but later fast growing. It attains 2-3 m height in 3 years.

Arjun yields upto 45 Kg. Dry bark chips on 3 year cycle without harming the tree, as bark is reproduced after 2 rainy seasons.

After care

Tending operations needed after planting are weeding, protection against fire, and frost for one or two years when forking takes place due to frost damage the weaker shoot is pruned off. Plants also need protection from browsing animals.

Medicinal uses

The bark is cooling, astringent, tonic, cardiac stimulant. It is a valuable remedy for heart diseases (Arjunarisht for sleeplessness). Also useful in bilious disorders, diarrhora, dysentery and haemorrhage. As an astringent the decoction is used for cleaning sores, ulcers, itches an ointment made of the bark with honey is efficacious for the cure of acne.

Other uses

Wood/Tiber wood is used for making agricultural implements, boat building, plywood, and cartwheel. The strong poles are suitable for providing support.

Fodder

Leaves contain 10.94% crude protein.

Silk production

Host for tasar silkworm.

Dye/Tannin

The bark, (24%) and fruits (10%) yield tannin, which is used for tanning and dyeing rawhide.

Oxalic acid

Bark also yields 14-20% oxalic acid.

BAHERA (*Terminalia bellirica*)

AJAY KHARE
Assistant Professor
Deptt. of Plant Physiology, JNKVV, Jabalpur (M.P.)

Botanical Name : *Terminalia beuerica*
Family : Combretaceae
Common Name : Bahera

Distribution

It is common throughout the greater part of India except arid regions in western Rajsthan, Punjab and Haryana.

Climate

Maximum temperature 36^0 - 46^0 C
Minimum temperature 1^0 - 15^0 C
Annual rainfall 900 - 3000 mm.

Soil

It is capable of growing on various types of soils provided with adequate moisture supply. The best growth is attained on deep sandy loam soils with good moisture supply and good subsoil drainage.

General description

A large, round crowned, handsome, deciduous tree. It grows upto 30-40 m. in height and upto 3 m. in girth, with a very straight tall bole, buttressed at the base. The bark is ashy gray with fine longitudinal crackes. Leaves are large, alternate, broadly elliptic, clustered at the end of branchlets, leaf sheding takes place from November to January, New leaves are copper red and appear between March to May.

Flowers are greenish white and appear alongwith new leaves. Fruit is drupe; sub-globose, very faintly five ridged when dry.

The tree is a light demander, fairly drought resistant, it is however very frost sensitive. It does not pollard well but is a good coppicer.

Propagation

Bahera is propagated by seeds.

Seed collection

Seeds are collected in winter from well-grown healthy trees. Freshly fallen ripe fruits are collected from the ground, swept clean in advance. The pulp is removed immediately and seeds are dried in the sun. About 176 to 420 seeds weigh one kilogram. The germinative capacity is 35-70%. Fresh seeds give best germination (which can reach upto 100%). About 5 Kg. Seeds are required to raise 1000 plants. Seeds retain viability for one year but their germinative capacity is reduced to 40-50%.

Nursery raising

Sowing of seeds in nursery is done in March-April in lines spaced at 20 cm. apart and seeds 5 cm. apart in lines prior to sowing seeds are soaked in water for 24 hours. Regular weeding and watering are nursery. Germination commences after 10 days and continues till 7 weeks when seedlings are 3-5 cm. tall, they are shifted to poly bags.

Direct sowing

Direct sowing is done in June-July in cultivated lines 3 m apart or in patches. The seeds are dibbled at 1 m. apart at 1.0 to 1.5 cm. depth and covered with soil.

Stumps

12-15 months old plants provide satisfactory stumps.

Planting

Planting is done in June-July when seedlings are 3-4 months old. Planting is done in the pits of 30 cm^3 size dug in advance. Spacing adopted may vary from 3 X 3 m. to 10-15 m. both ways.

As it is a tall growing tree, therefore it is suitable to grow as monoculture (Tree farming system). It can also be planted on farm boundaries. The tree has been grown in taungya plantations alongwith agricultural crops, which may be grown for 2-3 years between the lines of trees generally 3-4 m apart. Crops can be grown continuously if trees are wider spaced and are regularly pollarded for tasar silk worms.

Stump planting

Stump planting is done in July-August either in pits or in crowbar holes.

After care

The young plants should be provided with watering and weeding as and when required. Plantation should be protected from browsing animals.

Medicinal uses

The ripe dry fruit is astringent, tonic, and laxative. It is given in piles, diarrhoea, fever, dropsy, and leprosy, as well as in coughs. Its large dose becomes a narcotic poision.

Other uses

- Leaves are used for fodder.
- Tannin – The fruits and bark is used for tanning.
- Host for tusar silkworm.
- Twigs are used as firewood and for making charcoal calorific value is 5000 K.Cal./Kg
- Gum is obtained from wounded trunk, which is soluble in water but inflammable, and will burn like a candle.
- Kernels yield non-edible oil (38%) used for soap manufacturing.

HARAD (*Terminalia chebula*)

AJAY KHARE
Assistant Professor
Deptt. of Plant Physiology, JNKVV, Jabalpur (M.P.)

Botanical Name	:	*Terminalia chebula*
Family	:	Combretaceae
Common Name	:	Harad, Haritki, Harra

Distribution

It is found growing in U.P., H.P., Punjab, Maharashtra, M.P., Rajsthan Gujrat, TamilNadu, Karnataka, West Bengal, and Assam.

Climate

Maximum temperature	-	38° to 48° C
Minimum temperature	-	1° to 15° C
Annual rain fall	-	750 to 3250 mm

Soil

It is not exacting in its soil requirement and is found growing on clayey as well as sandy soils.

General description

It is a medium to large-sized deciduous tree with a short clean bole and a round spreading crown. The bark is dark brown, longitudinally fissured. The leaves are opposite, shortly petioled, ovate oblong, obtuse or cordate. The harad tree attains full size in 30 years. The flowers are small, whitish. The fruit is drupe, oval, glabrous. The time of flowering and fruiting is different due to edapho-climatic conditions. It may flower once or twice in a yrear. The new flush of leaves appears from March to May, while old leaves are shed from December to March. The fruits ripen from December to March.

Propagation

Planting out of nursery raised plants or stumps are more reliable than direct sowing because of the poor germinative power of the seeds, and high rodent damage.

Seed collection

Seed collection from forest floor and not from the trees is recommended Fruits attacked by insects are rejected. The fruits are dried before storage. About 140 to 220 dry fruits weigh one Kilogram The seeds store well for at least a year and germinate better than the fresh one.

Seed treatment

Soaking of fruit stone in cold water for 48 hours is recommended.

Nursery raising

Sowing in the nursery may be done in June - July. The fruit stones are sown after removing the hardened fleshy covering. Usually germination percentage of seeds is poor; therefore more seeds should be sown in the nursery.

Planting

About 12 to 15 months old nursery plants are planted out with balls of earth in the pits of 30x30x30 cm in size. Stumps perform better than planting of entire seedling. The best time for planting of harad is at the beginning of monsoon.

After care

The young plants are cared for by weeding and watering for at least one year because by that time plants will be well established. The seedlings and saplings also need protection from browsing animals.

Medicinal uses

Harad fruit is invaluable for preparation of various Ayurvedic medicines. Fruit is astringent, purgative, tonic, antidysentric, strengthens brain, enriches blood, antiparalytic, used in piles, cold, opthalmia, sore throat, dental care, bleeding and ulceration of gums, burns, diseases of spleen. An important constituent of trifala. Bark diuretic cardio tonic. Fruit soaked in water is a good cooling wash for eyes in conjuctivities. Kernel contains 40% oil.

Other uses

Fire wood	-	Twigs are used as firewood.
Fodder	-	Leaves are a medium quality fodder.
Timber	-	Timber is used for agricultural implements, carts, turnery, house-building etc.
Ink	-	The outer coat of harad fruit is mixed with Iron sulphate makes very durable ink.
Tannin	-	Root, bark, heartwood, sapwood, and leaves also contain tannin.

RATANJOT *(Jatropha curcas)*

V.K. GOUR
Department of Plant Breeding and Genetics
Jawaharlal Nehru Krishi Vishwavidyalaya, Jabalpur-482004, M.P.

Botanical Name : *Jatropha curcus*
Family : Euphorbiaceae
Common Name : Rajanjot, Vanarand

Introduction

***Jatropha curcas* L.** belonging to family Euphorbiaceae is a multipurpose shrub or small tree of significant importance because of its several industrial and medicinal uses. This species is one amongst the others, which got introduced in India approximately four centuries ago through Portuguese but remained commercially unutilized as a part of phyto-diversity away from the native country excepting its traditional use as folk medicine. It is the best example of a genera/species, which was known but remained unexploited/under exploited and the need of the time could make use it. The utilization of this species after a span of more then four centuries proves the need of plant genetic resource and justify conservation of biodiversity to fulfill the changing needs of the present and generations to come on the infinite time scale. It is now emerging as one of the most important crop with inherent multifold use, and is now becoming popular due to its oil, being used as **Bio-diesel**.

Jatropha oil is a potential source of "biofuel" in countries where the resources to import fossil fuel is poor. It could be served as a panacea for energy shortages in the future world. A renewable alternative source has been found which is being successfully grown and harvested as a biofuel in countries like Nicaragua. The seeds of the perennial shrub are used to extract oil, which on refining can be used as a biofuel in the form of extender to the High Speed Diesel (HSD) to the tune of 20%. The huge savings can be made in terms of not only money (both on spending and foreign exchange) but also on the non-renewable resource and decrease our dependency on foreign sources for oil. Processing also results in valuable by products like glycerine and bio manure in the form of cake.

Vernacular names

***Jatropha curcas* L.** is popularly known as Physic nut or purging nut. It is known by numerous vernacular names in the country. Sanskrit - Kananaeranda, Parvataranda; Hindi - Baghbherenda, Jangliarandi, Safedarand; Bengali - Bagbherenda, erandagachi; Marathi - Mogalierenda, ranayerandi; Gujrati - Jamalgota, ratanjota; Telagu - Nepalamu, peddanepalamu, adaviamidamu; Tamil - Kadalamamakku, Kattamanakku; Kannad - Adaluharalu, bettadaharalu, maraharalu, karnocchi; Malyalam - Kattavanakka, Kadalavanakka; Oria - Jahazigaba; Assam - Bongalibhotora; Garo Hills - Borbandong.

Origin and distribution

It is distributed in the tropical and sub-tropical parts of the world, chiefly in Africa and America. A number of scientists have attempted to define the origin of physic nut, but the source remains controversial. It is native of tropical America. It is highly probable that the nut is in Mexico (and Central America) since it is not found in these forms of vegetation in Africa and Asia but only in cultivated form. The "true" center of origin, however, still has to be found. It is assumed that the Portuguese brought the physic nut to Asia.

Today it is found scattered in all the tropical and sub-tropical regions of the world such as Brazil, Fiji, Honduras,

India, Jamaica, Panama, Puerto Rico and Salvador. It is known by more than 200 different names all over the world which indicates its great significance to man. Though it is found in almost all the state of India as a live fence. Gujrat, Rajsthan, Maharashtra, Madhya Pradesh and Tamil Nadu are the main states cultivating this crop.

Botanical description

Jatropha is large genus of herbs, shrubs and trees. The plant is reported to have been introduced into Asia and Africa by the Portuguese as an oil-yielding plant. It is found occurring almost throughout India and in Andaman Islands. It is found in a semi-wild condition in the vicinity of villages, mostly as hedge and fence. About 9 species have been recorded in India; some of them are grown in gardens for their ornamental foliage and flowers.

Linnaeus (1753) was the first to name the physic nut *Jatropha curcas L.* according to the binomial nomenclature. The synonymous names of the physic nut are:

Curcas purgans Medik
Ricinus americanus Miller
Castiglionia lobata Ruiz and Pavon
Jatropha edulis Cerv. Gaz.
J. acerifolia Salisb.
Ricinus jarak Thunb.
Curcas adansoni Endl.
Curcas indica A.Rich.
Jatropha yuctanensis Briq.
Curcas curcas (L.) Briton and Millsp.

The taxonomical hierarchy of Jatropha is given as under after revision by Dehgan and Webster (1979):

Tribe - Joannesieae of Crotonoideae
Family - Euphorbiaceae
Genus - Jatropha

Subgenera - * Jatropha
* Curcas (10 sections & 10 subsections)
Species - Approx. 170 sp.

The genus name *Jatropha* derives from the greek *iatros (doctor)* and *trophe (food)* which implies medicinal uses.

Jatropha curcas L.

A large shrub, 3-4 m. high, leaves alternate, 10-15 cm. x 7.5 – 12.5 cm; deciduous, alternate but apically crowded, broadly ovate, acute to acuminate basally cordate, usually palmately 3 or 5-lobed, glabrous; flowers in April-May in several to many greenish loose panicles of cymes, yellowish green, *c.* 7 mm across; Male flowers many with 10 stamens, 5 united at the base only, 5 united into a column. Female flowers borne singly with elliptic 3 celled, triovulate ovary with 3 spreading bifurcate stigma; fruits *c.* 2.5 cm. long, ovoid, black, breaking into three 2-valved cocci; seeds ovoid-oblong, dull brownish black. Seed resembles castorseed in shape but is smaller in size (wt. 0.5-0.7 g.; length 1-2 cm.) and of dark brown colour.

All parts of the plant exude a sticky, oplascent, acrid and astringent latex, containing resinous substances (14.6% in the latex coagulum) but no rubber. The latex dries to a bright reddish brown, brittle substance, resembling shellac; it stains linen indelibly and can be used as marking ink. The bark contains tannin (37%, dry basis). It also contains wax, resin, saponins, reducing sugars and traces of a volatile oil. The bark yields a dark blue dye reported to be used in the Philippines for colouring cloth, fishing nets and lines. A dye may be extracted also from the leaves and tender stem, and concentrated to yellowish syrup or dried to a blackish brown lumpy mass. The dye imparts to cotton different shades of tan and brown which are fairly fast.

Other species that belongs to section curcas are: *J. pseudo-curcas* Muel. Arg., *J. afrocurcas* Pax, J. macrophylla* Pax* and Hoffm, *J. villosa#* Wight (Syn. J. wightiana

Muell.Arg), *j. hinotonii* Wilbur., *J bartlettii* Wilbur., *J.mcvaughii* Dehgan and Webster., ***J. yuctanensis*** Briq. (**Syn. *J. curcas***). The species marked with asterisks are of east African origin whereas, with# mark is of Indian origin. Most of the Jatropha species are native to the new world; however approximately 66 species are native to the old world. *J. Multifida* L. and *J. podagrica* Hook of section *Peltatae,* J. integerrima of section *Polymorphae,* and *J. gossipyfolia* of section *Jatropha* are well known and cultivated throughout the tropics as ornamental plants.

J. glandulifera var. green Roxb (Jangli arand)

It is a weed found in southern as well as in central part of India. Leaves and stems are green in colour and possess oil glands. It often confuses with *J. gossipyfolia, which* is a more common plant. *J. glandulifera* is distinguished from *J. gossipyfolia* by the serrate gland-tipped leaves, long branched gland-tipped stipules and greenish yellow flowers. The seeds contain 20-22% oil.

J. glandulifera var. Pink Roxb (Pink Jangli arand)

It is awed plant and it differs with previous species by its pinkish leaves and stem.

J. multifida (Coral plant/Bhadradanti)

It is an ornamental shrub, 2-3 m high; leaves are 5-10 lobed, flowers are coral red in colour, borne on many flowered, long pedunculate, flat-topped, terminal cymes.

J. gossypifolia var. rose (Bherenda)

It is profusely flowering ornamental shrub, flowers are dark red or crimson, capsules are small and possess less oil. Seeds are grayish red with a caruncle.

J. gossipyfolia var. pink (Pink Bherenda)

It is also a pink flowering shrub with small capsules and rest other features similar to previous species.

J. nana Dehz or Gibs

It is a small shrub, leaves entire or three lobed, flowers pedicellate, in few-flowered terminal paniculate cymes fruits are small, oil content is less.

J. podagrica Hook. (Gouty stemmed Jatropha)

It is native of America. It is a bonsai plant, flowers are rose, leaves broadly ovate, alternate, and the basal stem is bulged like a bottle. Seeds propagate it.

J. tanjorensis Ellis and Saroja

It is found abundantly in Tanjore, Pudukottai, Trichirapalli and Ramnad districts of Tamil Nadu, India and grown as fence. Plant showed intermediacy in phenotypic characters of *J. curcas* and *J. gossypifolia* L. It has been confirmed cytologically and biochemically as a natural interspecific hybrid between these two species.

J. integerrima Jacq. *Syn.* J. panduraefolia *Andr. (Fiddle-leaved Jatropha)*

It is native of America. It is widely grown in Indian gardens as ornament. Leaves arranged in spiral, obovate dark green with green-brown pigmentation on the lower surface, short petiole with dark pigmentation, large crimson red unisexual flowers with 8-10 red anthers arranged in two whorls. It is drought tolerant ornamental with bright crimson red flowers. Its seed oil is rich in linoleic acid. It has high degree of resistance to leaf eating caterpillars. Seeds propagate it.

Varieties

Two types viz., (i) Nicaragua types and (ii) Mexican types are reported in the Western world. Mexican types are free of toxic substances and the Nicaragua types are poisonous types. The centers of diversity of land races and ecotypes are central and south America. In Madagascar *J. mahafalensis* (2n+22) is commonly cultivated. In India as there is no named varieties the local types are widely cultivated.

Climatic requirements

It is hardy to dry weather conditions and can be grown over a wide range of arid or semi-arid climatic conditions. Whilst Jatropha grows well in low rainfall conditions (200 mm), it can also tolerate to high rainfall (upto 1200 mm) conditions. It can stand long periods of drought by shedding most of its leaves to reduce the transpiration loss. It tolerates annual temperature range of 18-28° C. For the emergence of seeds, hot and humid climate is preferred. The flowering is induces in rainy season and bears fruits in winter.

Soil and Soil preparation

Jatropha curcas is a hardy plant well adapted to arid land semi arid conditions. It has low fertility and moisture demand and can come up stony, gravelly or shallow and even calcareous soils. The crop is undemanding in soil type and even does not require tillage. It does not thrive in wetland conditions.

Propagation

Seeds or cuttings rapidly and easily propagate it. The seeds can be sorted and the plumpy and filled seeds alone to be selected. The overnight soaking of seeds in water improves the germination percentage. The pot mixture may be filled up in 18 x 13 cm poly bags and the seeds are sown. The seedlings germinate within a week time and will be ready for transplantibg in 45 days. Plants from seeds develop a typical taproot and four lateral roots. Seedlings start bearing from 3-4 years. The rooted semihard and hard woodcuttings also may be used for planting. Cuttings are ready for planting in 2-3 months time grows rapidly and they start bearing in the same year of planting. It can be cut or lopped at any desired height and is well adapted for hedges. The general observations and literature reveals that it flowers in hot and rainy season and sets fruit in winter when it is leafless.

Seed rate and spacing

The seed rate for planting one hectare is 5-6 kg. The spacing is 2 x 2 m and will accommodate 2500 plants per hectare under irrigated or partially irrigated conditions. In south Indian conditions 2 x 1.5 m spacing (3250 plants) is found to be ideal. On rainfed wasteland, high density planting with a spacing of 2 x 1 m or 1.5 x 1.5 m accommodating 5000 or 4444 plants per hectare respectively shall be desirable.

Field Preparation and Planting

The land should be ploughed once or twice depending on the nature of soil. In case of heavy soil deep ploughing is given whereas in light soils, shallow ploughing is enough. It loose soil pits are dug at 30 x 30 x 30 cm size and filled with soil and organic mix. Under clay soil the trenches may be taken with 30 cm depth, 30 cm width and 2 m distance between them. Pits are filled with soil and compost or organic manure @ 400g/pit.

a) Direct Planting

The seeds or cuttings are directly planted in the main field with the onset of monsoon during June-August months. For direct sowing two seeds are dibbed at each spot at the specified spacing. When the seedlings are 4 weeks old, weaker seedlings should be removed to retain one healthy seedling on each spot and the seedlings so removed could be used for gap filling. The large semi hard wood or hard woodcuttings can be directly planted in the field.

b) Transplanting

For transplanting seedlings are grown in polybags ½ kg. Capacity filled with soil and organic manure mixture @ 100g per polybag plus 400 g soil. Two seeds should be sown around 6 cm deep in polybag and watering should be done regularly. When the seedlings are around 4 weeks, weaker of the two seedlings should be removed and used for gap filling. The grown up seedlings or cuttings are transplanted in the main field. The pits are filled with 500 g farmyard manure and 100 g neem cake to ensure profuse rooting.

Crop management

Crop architecture plays an important role in a monoecious plant like Jatropha where in proper pruning will help in producing more flowers and thereby enhance the seed setting probability and ultimately the yield. The preliminary experiments reveal that the pruning at 1.5 m height is ideal to manage. The pinching of terminals is essential at six months age to induce laterals. The application of GA (100 ppm) spray induces flowering and pod development. The entire plant has to be cut to ground level leaving 45 cm stump once in 10 years. The regrowth is quick and start yielding in about a year.

Manure and fertilizers

Although Jatropha is adapted to low fertility site and alkaline soils, better yield is obtained on poor quality soils if fertilizers with small amounts of calcium, magnesium and sulphur are used. Mycorrhizal associations have been observed with Jatropha and are known to aid the plants growth under conditions where phosphate is limiting. In general application of super phosphate @ 150 kg/ha and alternated with one dose of 40:100:40 kg/ha NPK at 6 monthly intervals improves the yield. The application may coincide with rainy season or ensure proper irrigation immediately after application of fertilizers. From fourth year onwards 10% extra super phosphate should be added to the above dose.

Irrigation

During the dry period, the crop is irrigated 7-15 days interval depending on the requirement. Though the weekly irrigation is preferable, fortnight interval is compulsory. Drip irrigation is not ideal as it induces too much of vegetative growth.

Inter culture operations

The field should be kept free from weeds at all times. Around 3-4 weedings in the initial period are enough to keep the field free from weeds until the crop crosses the stage of grand growth period. Light harrowings are beneficial during the early growth stage.

Intercropping

Jatropha can be intercropped with other fruit plantations till the main plantation comes to fruiting. The need based field crops suiting to the agro-climatic conditions should be grown to harvest substantial as intercrop. In pure Jatropha plantations herbals, pulses or oilseed etc. can be grown as intercrop during initial growth period.

The agronomists are finding appropriate field crop, medicinal and aromatic plants as intercrop to add to farmers' income till the plantation attains reproductive phase.

Harvesting

The seeds are harvested as needed for medicinal purpose. For oil purpose the seeds are harvested when the fruits are turning yellow. The pods are collected manually and seeds are separated mechanically or manually. The seeds are to be dried in sun for four days (6-10% moisture level) before packing.

Yield and Production

The correct yield estimates for this TBO is not available. With good care every year the expected average production of seeds (Table 2) touches 5 q./ha of seed harvest from 4th year and from 6th year onwards it is substantial.

Year of Planting	Expected yield/ha of irrigated crop (Kg)
1st	250
2nd	1,000
3rd	2,500
4th	5,000
5th	8,000
6th onwards	10,000

Few reports reveal that the grown up Jatropha plants yield 4-6 kg dried seeds per plant with an estimated yield of 10-12 t of seeds per ha in irrigated plantations from 6th year onwards. With the minimum market price of Rs. 5/- kg of seeds a gross return of Rs. 60,000/- per ha. Per annum could be realized.

Post harvest processing

The dried seeds contain 66% kernel, which can yield between 46-58% oil. The extraction process standardized for edible oils is followed for extraction of Jatropha oil also. The oil is extracted by solvent extraction or mechanically by hydraulic or screw press.

Using a well-established procedure, manufacturers can mix the oil with methanol in a particular proportion at a specific temperature and separate the glycerol content, stirring continuously for 2 hours. Washing twice the residual solution yields purified bio-diesel. The production unit comprises, among other things, a vessel with a heater, a stirrer assembly, a container for methanol and oil and few setting tanks to separate glycerol and a washing tank to purify the fuel.

Chemical constituents

The seeds contain starch, dextrose, gluten, a free acid and an active lipase are present. Analysis of the seed gave the following values:

Moisture(%)	06.62
Protein(%)	18.20
Fat(%)	38.00
Carbohydrate(%)	17.98
Fibre(%)	15.50 and
Ash(%)	04.50

The seed possess poisonous and purgative properties but they are rarely used as purgative. Three to five seeds, slightly roasted and decorticated, are sufficient for active cathartic; it seldom produces nausea and vomiting, but cause a sense of burning in the stomach. They contain two toxic principles, curcine or curcasin, a toxalbumin resembling ricine, and a resinous substance (possibly resinolipoid) with nauseating and purgative action.

The kernel forms 60-68% of the weight of seed and yields an oil (46-58% of kernel wt.; i.e. 30-40% of seed wt.) which when fresh is practically colourless and odourless. On standing the oil acquires a pale yellow or yellowish brown colour and a disagreeable odour. The oil is obtained from decorticated seeds by expression or solvent extraction and is known in the trade as Curcas Oil. The oil is semi-drying and may be employed for the preparation of non-or semi-drying alkyds.

Plant protection

Pests

Bark eater and pod borer are the major pests. They can be controlled with herbal pesticides such as mixtures of vitex, neem, aloe, *Calatropis* or rogor @ 2 ml/lit. of water.

Diseases

Collar rot may be the problem in the beginning and can be controlled with 0.2% COC.

S. No.	Plant Part/ Oil	Process*	Uses*	Name of the place/country
India:				
1	Seeds	Fried and powered taken with molasses	Stomach ache	Travencore
2	Seeds	-	Antidote for poisoning	Travencore
3	Oil	-	Illuminant	
4	Oil	-	Making soaps and candles	
5	Oil	External application	Skin diseases and Rheumatism	
6	Oil	Abortifacient	Efficacious in dropsy, Sciatica and Paralysis	
7	Tender twigs	-	Cleaning teeth	
8	Leaves	-	Leaves are used as feed for eri silk worm	Assam
9	Leaves	-	Rubefacient, galactagouge, suppurative and insecticidal properties, useful in foul ulcers, tumours and scabies	
10	Juice	-	Relieve toothache and strengthen gums; external application for piles	
11	Latex/ Twig sap	-	Considered styptic, purgative and haemostatic and used for dressing wounds and ulcers	
12	Sap	Emulsion of the sap with benzyl benzoate	Considered effective against scabies, wet eczema and dermatitis	
13	Leaves and Roots	Decoction	Given for diarrhoea, used as antidote for snake bites	
14	Roots	Contains Yellow oil	Strong anthelminthic action	
15	Bark	Bark rubbed with asafoetida and buttermilk	The paste so obtained is used in case of dyspepsia and diarrhoea	Konkan
16	Bark	Decoction	Rheumatism and leprosy	
Abroad:				
1	Seeds	-	Anthelmintic	Brazil
2	Seeds	Ground Seeds with palm oil	Rat poison	Gabon
3	Oil	Boiling oil w ith iron oxide	Varnish	China
4		-	Wool spinning	England
5	Oil	-	Adulterant of Groundnut oil	Senegal
6	Oil	Applied to hair	Growth stimulant	Java
7	Leaves	-	Tender leaves are eaten after cooking	Java and Malaya
8	Leaves	-	Fumigate houses against bed-bugs	Ghana
9	Juice	-	Used as purgative and haemostatic	Java
10	Juice	-	Stupefying fish	Philippines
11	Bark	Yields a dark blue dye	Colouring cloths and fishing nets	Philippines

** Source of information: Wealth of India*

Uses

Medicinal Uses

The Jatropha plant and its various parts are used for varied purposes with traditional knowledge as folk medicine in the India and abroad as well, is listed below. The various plant parts, its uses and the brief processes available is given.

Industrial and miscellaneous uses

The oil is being extensively used in wool spinning, cosmetics, soap, varnishes and candles making. The oil is used as an illuminant because it burns without emitting smoke and is being used as a lubricant also.

Oil in Plant Protection

The crude Jatropha oil has been used as safer material to control stored food grain insects, snail, housefly, aphids, pink boll worm, Spodoptera, Helicoverpa, Damping off, Wilt etc. The fumes are used against house bed bugs.

Oil Cake

The oil cake contains rich source of nitrogen (3.2-4.4%), Phosphorous (1.4-2.09) and potassium (1.2-1.68%) and valued as organic manure. Though the cake contains 38% proteins, it is non-edible due to the presence of toxic substances and could be converted into cattle feed through latest technologies. The cake protein may be employed as a raw material for making plastic and synthetic fibres.

Glycerin

The bi-product glycerin emanating out of bio-diesel clarification is used for soap making.

Biogas

It is produced from the oil cake and it contains 70% methane. Its degradation rate is 70-80%.

Feed

The leaves can be used to rear Eri-silkworm and Tusser silkworm. In Java and Malaysia the tender leaves are eaten after cooking.

Dye

The bark yields a dark blue dye and it is used for colouring cloth, fishing nets and linens. The yellow dye extracted from leaves and tender stems is used to colour cotton cloths.

Soil enrichmen

The Jatropha leaves provide plentiful organic matter and increase the microbial activity including earthworm. The tender branches and leaves are widely used to prepare manure. The tree releases the O2 into the environment and accumulates CO2 in the plants. On incorporation it enriches the soil carbon. The plants reduce soil erosion and will help to conserve moisture.

Wasteland Development

It can be grown as agro-forestry crop on wastelands or barren and marginal lands where no irrigation facilities are available. It can also be used for land reclamation. It is a crop with low capital investment, short gestation period and long productive period. The country's 175m ha wasteland could be put effectively for rural income generation and fuel production.

Fence Crop

It is commonly grown as a live hedge around agricultural fields as it can be easily propagated, grows rapidly, hardy and is not browsed by goats or cattle.

Ornamental Crop

Nine out of twelve species found in India are being grown in garden for their ornamental foliage and flowers.

NEEM (*Azadirachta indica* A. Juss)

SHARAD NEMA
Research Associate
Deptt. of Plant Physiology, JNKVV, Jabalpur (M.P.)

Botanical Name	:	*Azadirachta indica*
Family	:	Meliaceae
Common name	:	Neem tree, Margosa tree, Indian Lilac Prabhadrah (Sanskrit) Nim, Nimb (Hindi)

Uses

Neem, *Azadirachta indica* A. Juss (Syn: *Melia indica* Brand and M. *azadirachta* Linn.), is one of the most valuable and yet the least exploited of tropical trees. It has a multiple uses such as firewood, timber and in pharmaceutical and entomological preparations. Neem tree has great potential for agriculture, industrial and commercial exploitation. Each and every part of the tree from its roots, trunk, bark, leaves, flowers, fruits, seeds, sap and gum are known to have some use and have a place in the traditional folklore and medicine (especially in the Ayurvedic and Unani systems of medicine), so much so it almost plays the role of a village dispensary in this land.

The leaves contain minerals like calcium, magnesium, iron and phosphorus, protein, vitamin A, crude fibre, carbohydrates, fat and alkaloids like nimbosterol, azadirachtin, nimboflavone and nomicinol. Medicinally, its leaves are used to cure many diseases of the bladder, kidney, eyes, and skin, for the explusion of intestinal worms, and with honey for curing jaundice, boils and urticaria.

The flowers contain nimbosterol, fatty acids and a highly pungent essential oil. The dry flowers are considered as a stimulant, tonic and stomachic. The flowers are useful in some cases of atonic dyspepsia and general debility. The bitter oil extracted from the flowers is useful in skin diseases.

The bark contains a resin, nimbin, nimbinin, nimbidin, essential oil, tannins, the bitter principle margosine, desa-acetyl nimbin, inorganic phosphorus, sulphur and iron. The bark is a bitter tonic, astringent and antiperiodic. The bark preparations are useful in fever, thirst, nausea, vomiting, skin diseases, scorpion and snakebites. The oil of the bark possesses a marked antibacterial spectrum against Gramve and Gram +ve bacteria. The fresh, tender twigs are used to clean the teeth, particularly in cases of pyorrhoea.

The heartwood of the tree is slightly aromatic and is reported to yield nimatone, tannins, inorganic calcium, potassium and iron salt.

The pericarp of the fruit is rich in carbohydrates and constitutes a promising substrate in methane gas generation. The seeds contain calcium, magnesium, phosphorus, an amorphous morgosopicrin, diacetylnimbin, thionemone, azadirachtin, salanin etc. nimbinine, nimbanidol, nomolicinol, and salannotide. The seed-kernel contains about 45% of fixed oil called neem oil. The oil is non-drying, bitter and greenish-yellow to brown in colour and has a strong, characteristic, garlic-like odour. The bitter principles of the oil are nimbidin, nimbin and nimbidol. The other active principles of the oil are salannin, meliantriol, azadirachtin, azadiradione, azadirone, melianone, nimbolin, salannol, vepinin and volasinin. The oil displays a stimulative, antiseptic, an alternative effect when used for massage of the body and is said to possess anti-inflammatory properties. Medicinally, the oil is employed as an antihelminthic, useful in bleeding gums, pyorrhoea and asthma. It is used externally as a stimulant in convulsions and collapse from fevers, cholera and fits. Several pharmaceutical preparations like liquors, emulsions, ointments, liniments, as well as medicinal cosmetics like creams, lotions, soaps shampoos, hair-tonics, toothpastes and gargles have been prepared with nimbidin and with the salts of the acid derived from it. Nimbidin is reported to be efficacious in the treatment of skin diseases such as warts, extensive and persistent dandruff, furunculosis, arsenical dermatitis, scabies and saborrhoeic dermatitis. Nimbidin is also known to exhibit pronounced anticholinergic, diuretic, antihistamine, antinicotinic, anti-ulcer and moderately sedative effect in experimental animals. The oil is used in manufacturing a good number of insecticides as well.

Neem-cake is obtained after expressing the oil from the seeds. The cake extensively used in crop production as organic manure.

The gum exudes from the bark in the form of clear, bright, amber-coloured tears or fragments. The gum is prized medicinally as a stimulant, demulscent and tonic.

Distribution

Neem is believed to be a native of upper Burma and possibly of the Siwalik, Deccan and other parts of South India. It is widely distributed in tropical Africa, Australia, the Andamans, Pakistan, Sri Lanka, Myanmar, Malaysia, Indonesia, Bangladesh, Thailand, Middle East, Sudan, Nigeria, Fiji, Mauritius, Central and South America, the Carribean, Puerto Rico, the Virgin Islands, Haiti -and other places. In India, it is cultivated all over. It grows wild in the Siwalik Hills, in the dry forests of Andhra Pradesh, Tamil Nadu and Karnataka. The tree is well suited for afforestation.

Description of the plant

Neem is a large glabrous tree with a straight trunk. The branches spread widely and form a oval crown.

The leaves are alternate, imparipinnate, medium long, the leaflets are 9 to 13 in number, nearly opposite, shortly petiolate.

The inflorescence is an axillary panicle. The flowers are white, with a strong smell of honey, especially at night, pentamerous, on short slender, pedicles with short, scattered hairs. The flowers appear from March-May and the fruit ripen during June-August.

Soil

Neem grows even in a nutrient deficient soil. It grows well on moist, dry, stony, clayey or shallow soils, including moderately saline and alkaline soils, compact clays and lateritic crusts. However, silty flats, clayey depressions and land prone to inundation are not conducive to its growth. It can tolerate a soil pH of up to 10.

Climate

The tree is very hardy and can survive high temperatures and altitudes between 50 and 100 m and little rainfall, as Iow as 130 mm per year, and long stretches of drought. In India, it thrives best in parts where normal rainfall ranges between 480 to 1 000 mm and maximum temperatures as high as 48°c.

Cultivation

Propagation

Seed propagation is the common method of growing neem trees. It can also be propagated easily by vegetative means, like root-and shoot cuttings. 'Root- and shoot-cuttings can be made from 12-13 month old nursery plants. A success rate of 53% from root-shoot-cuttings is reported from Maharashtra.

Nursery raising

The seeds are collected from June-August, which coincides with the rainy season. The seeds remain viable for a short period of 3 to 5 weeks only. Hence, they should be sown immediately or within 2-3 weeks of collection. The seeds can remain viable up to 2 months when stored in airtight tins, in pots with wood-ash and in gunny bags. The seeds are depulped before sowing.

The nursery is raised in beds of rows 15 cm apart with the seeds 2.5 cm apart in a row. The beds are sparingly watered after sowing. The seedlings are transplanted from when they are about 2 months till l-year-old. It is preferable to plant the seedlings when they are 7-10 cm in height. The nursery is protected against frost by providing screens. The seedlings can be raised in baskets and polythene bags also.

Planting

Direct sowing is easy and can be successfully done by dibbling, broadcast, sowing in lines or in trenches, depending on the site conditions.

Stumps from 2-year-old plants have been reported to give a better survival and faster growth rate than those from 1-year-old plants. The stumps are prepared with 2.5 cm of shoot and 23 cm of root pieces and should be planted at the onset of the rains.

Neem in Agroforestry systems

Neem is well suited for arid zones in India. It is the least affected by automobile exhaust and is best suited for roadside planting. It is also well suited for afforestation of wastelands and for planting under agroforestry systems.

Neem is interplanted with groundnut, bean and millets in farmlands. It is reported that neem trees can be planted in highly alkaline soils if the land is levelled, bunded and gypsum applied. The trees can then be grown along with agricultural crops.

Pests and Diseases

The plant is not attacked by any disease of a serious nature.

Harvesting and Yield

The fruits mature in June-July and produce about 50 kg/tree of berries.

Distribution

India, Africa, Central and South America, Andamans, Australia, Bangladesh, Carribean Fiji, Indonesia, Haiti, Malaysia, Mauritius, Myanmar, Nigeria, Pakistan, Puerto Rico, Sri Lanka, the Middle East, Sudan, Thailand and Virgin Islands.

Mulahati or Liquorice *(Glycyrrhiza glabra Linn.)*

Dr. A.S. Gontia
Principal Scientist,
Deptt. of Plant Physiology, JNKVV, Jabalpur (M.P.)

Botanical Name	:	*Glycyrrhiza glabra Linn.*
Family	:	Papilionaceae
Common name	:	Mulanati

Mulahati is a hardy perennial herb and under shrub, belonging to the family papilionaceac. As its roots are very seat, it is popularly known a liquorice. In Greek glycyrrhiza means sweet roots.

In India it is known as Jethimad, Yasthimadhu, Jesthimadhu or mulahati. It has been used in pharmacy for thousands of years as an important medicinal plant. It is a principal drug of susruta. The Egyptians, Greek and Romans utilized it for medicinal purpose.

The dried, peeled or unpeeled underground stems, and roots constitute the drug. Its extract contains about 5-20% sapnin, the glycoside glycyrrhizin, which has healing effect on peptic ulcers. It also contains coumarine compounds, herniarin and umbelliferone and the flavors liquiritin, liquiritigenin, isoliquiritin and isoliquiritigenin. The glycosides liquiritoside and isoliquiritoside are also present. Its main uses are demulcent, mildly laxative, expectorant and it is an important constituent of all cough and catarrh syrups, throat lozenges and pastilles. It is also known for its analgesic, anti-inflammatory and anticonvulsive properties. The extract is reported to land sparkle and aroma to beer.

Origin and Distribution

Liquorice is distributed in the subtropical and warm temperate regions, particularly in western china, parts of Asia, Persia, the Asian Republics of the erstwhile USSR and Afghanistan. It is also distributed in the warmer parts of the Mediterranean region of North Africa, Spain, Italy, Yugoslavia, Greece, Syria, Hungary and South Russia, Southern Europe to Pakistan and Northern India are considered its place of origin. In India it is mainly cultivated in Baramulla, Srinagar, Jammu, Dehradun, Delhi, Anand, Indore and Bangalore.

Description of the plant

It is a perennial herb or under shurb attaining a height of 1-2 m with long cylindrical, burrowing rootstock and horizontal creeping stolens which are 1.5-1.8 m in length. The leaves are alternate, pinnate with 9-17 ovate, yellowish green leaflets of 2.5-5.0 cm. long. The flowers are borne in erect, axilliary long spikes usually 10-15 cm long, lavender to purple in colour and 1.25 cm long. The seedpods are maroon, 3 cm long, oblong, pointed, flattened contain 2-4 kidney-shaped seeds. The chromosome no. Is 2n = 16 and plant is self-pollinated. The root system is highly branched and confined to the upper crust of the soil, it consists of a short taproot with a large number of rhizomes.

Species and Varieties

1. *Glycyrrhiza glabra* var. typical regal and herd. This is called Spanish liquorice. It attains a height of 120-150 cm and bears papilionaceous flowers of purple blue colour. It fetches a higher price in the market and is the sweetest of all.

2. *Glycyrrhiza glabra* var. glandulifera, waldest and kit. It is called Russian liquorice. It is rich in glycyrrhizin (10-14%) but the sweet test is accompanied by a perceptible but wild bitterness due to presence of liquiritin and isoliquiritin.

3. *Glycyrrhiza glabra* var. violoceae Boise. This is called Persian liquorice bears violet flowers. The plant produces thick, hard and lumpy roots and is considered inferior in quality.

Haryana Agriculture University has released a variety Haryana Mulahati No. I, which is a Russian collection. It has broad, long and dark green leaves and produces approximately 200 cm long roots at harvest. This variety is resistant to leaf spot disease and gives 70-80 q/ha of roots in 3 years with 7.5% glycyrrhizic acid. An accession, EC 21950 has been identified at the all India coordinated research project on Medicinal and Aromatic plants, which is said to be tolerant to the root-rot disease.

Soil

Loamy soils of light texture, with a pH of 6-8.2 are ideally suited for economic and biological productivity. Deep moist soils, subjected to periodical inundation are also suitable. This crop can withstand salinity.

Climate

It prefers a warm and dry, subtropical climate with well-defined winters. Heavy rains or frost reduce its growth. 50-100 cm. annual rainfall, supported with irrigation is most appropriate for its proper growth.

Cultivation

Seed Propagation

Though plants can be grown from seeds, this is seldom practiced because of germination problem. Seed treatment with 0.0025 to 0.005% of succinic acid combined with scarification (in sand) was found to enhance the germinability. The treated seeds are drill sown 15 cm apart in raised nursery beds (1 x 3 m) that are well prepared and manured. The germination starts after 15 days and is completed with in a month. The seedlings are pricked out when they are 10-15 cm. high and planted in the main field.

Clonal Propagation

This is generally excercised in July or august. It can also be done in February-March if the irrigation facilities are adequate. In this method roots are used as clones. Root cuttings of 6-9 inches are used. During the planting 3/4 part of it is kept inside the soil white ¼ part is exposed outside soil. The plants are also produced from runners or underground stems which are prepared in to cuttings of 6-9 inches long. The cuttings are kept in moist sphagnum moss for about 8-10 days and afterwards, when their buds start sprouting; they are planted in the main field.

Preparation of main field and planting

Field should be well prepared by ploughing, harrowing and then levelled. The planting ridges 70 cm wide and 45 cm apart are marked out. The planting should be done while maintaining the distances of 75 x 45 cm. between plants and rows, respectively. The crowns are covered with 5-7 cm of wet soil. Treating the cuttings with Ceredox B-2 before planting results in good crop establishment. In order to obtain good growth and development of roots, the field must be liberally manured. Compost or Fym @10-15 t/ha is enough to meet the requirements. Besides a fertilizer doses of 60 : 40 : 20 NPK/ha per year are recommended for application in the form of urea, super phosphate and murate of potash. Of these a full dose of super phosphate and potash are applied as a basal dose at the time of planting, whereas nitrogen is applied in three split doses i.e. planting, 6 months later and one year of growth.

Irrigation

Regular irrigation is necessary unless cuttings put down roots and establish themselves. Later, the watering may be done at intervals of 8-10 days for 5-6 months and afterwards as and when required. Once the plants are established, they need no special attention, as the crop is fairly drought resistant.

Pruning

Pruning of branches at the time of flowering enhances the root quality.

Intercropping and Weeding

Since the crop occupies the land for a longer period intercropping of potato, carrots or the cabbage may be carried out in the 1st year to get more output from the land. The field may be hoed periodically to keep the field weed free.

Harvesting

The crop is ready for uprooting in 2½ - 3 years after planting and just before the plants have borne fruit. Generally the roots are collected after the rainy season is over. A trench of about 60 cm deep is first dug at the side of the ridges then, by working inwards the soil is loosened form the roots so that they can be pulled out easily. The broken root parts are left in the soil and new sprouts are thrown out in spring and, for raising the succeeding crop, only the gaps need to be filled with rooted cuttings.

Processing

Roots and underground stems are cut in to 15-20 cm long pieces, 1-2 cm in diameter and are dried alternately in shade and sun, this may take several months. The drying process reduces the weight to 50% and the moisture from 50-60%. Artificial drying can also be done at 30-40° C by using mechanical driers.

Pest and Diseases

Pests – Termites cause a lot of damage to young plants during the dry period. Chloropyriphos @ 4ml/litre can be given as a soil drenching. Cotton ash weevil (*Myllocerus undecimpunctatus*) is reported to cause injury to leaves and this insect can b controlled by spraying metasystox (1.5 ml/litre).

Disease – The crop is found to be infected by root-rot disease. The roots and stolens of the affected plants turn soft and pulpy. Drenching the soil with 4: 4: 50 Bordeaux mixture is done during February to control the disease effectively. Rhizoctonia spp., Sclerotium spp. and Fusarium spp. cause root rot, collar rot and wilt diseases, respectively. To control the disease it is recommended to follow the cultural practices viz : crop rotation, summer ploughing, well drainage. *Cercospora* spp. is reported to cause leaf spot disease. Spray of 0.05% Bavistine was found effective. However, one protective spray should be given one and a half to two months after planting. The other fungi, which are reported to cause leaf spot diseases, are Alternaria tenuis, Septoria glycyrrhizae or Phyllosticta glycyrrhizae. Uromyces glycyrrhizae causes leaf rust. Leaf blight is another serious infection, which can be controlled by foliar spraying of Blitox (0.2%) 3-4 times, at an interval of 5 days.

SARPAGANDA (*Rouvolfia serpentina*)

A.B. TIWARI
Associate Professor
Deptt. of Plant Physiology, JNKVV, Jabalpur (M.P.)

Botanical Name	:	*Rouvolfia serpentina*
Family	:	Apocynaceae
Hindi	:	Chotachand, Chotachandrika
Bangali	:	Chandra, Pagal ki Butee
Sanskrit	:	Sarpaganda
Marathi	:	Hrkaya
Telgu	:	Patala-gandhi.

Introduction

Rauvolfia is a plant (genus) under Apocynaceae family popularly called as sarpagandha, or chandrika in Hindi and Sanskrit. The trade name Rauvolfia refers to a 16th century German botanist and physician Leonard Rauvolf. Serpentina refers to long tapering snake-like roots. The roots of the plants were used for centuries by the Ayurvedic and Unani system of medicine in India for various type of ailments, however its importance in modern medicine was recognized in 1952, only after the isolation of a pharmacodynamic principle, reserpine, in the roots of the plant which revolutionzed the therpeutic use of the drug as anti-hypertensive and sedative.

It is a small, erect, evergreen perennating, undershrub, 15-45 cm (rarely 90 cm) high. Taproot tuberous soft, sometimes irregularly nodular, bark pale brown, corky with irregular longitudinal tissues, rarely lenticellate. Leaves are thin, large, elliptic- lanceolate or obovate in shape, acute or acuminate arranged in whorls of 3-4, dark green above, pale green below. Petioles 3 mm long and somewhat obscure owing to the blade running down into the petiole.

Flowers are white or pinkish, often tinged with violet, small and 2-4 per cyme, are born on red stalks. These salver shaped corolla tubes are purplish black, when fully ripe, 1-2 seeded and shape of seed is ovoid. Albumen is fleshy. The intensely bitter root is long, thick and round on the upper side and gradually tapers downwards. Its diameter at the thickest portion varies from 1.2 to 2.5 cm.

Distribution

Rauvolfia species found in the tropical and subtropical parts of world, five are recorded from India. However, two of these i.e. *R. sepentina* and *R. tetraphylla* form the commercial source of resperine. Besides resperine, other important alkaloids have been isolated from roots. Different varieties of Rauvolfia have been found growing in Central South America, Africa, India, Java, Sumatra, Thailand, Bangladesh, Burma, Malaysia and Sri Lanka.

Climate

The Plant is known to grow in a wide range of climatic conditions. However, it flourishes better in hot and humid tropical areas and can be grown in the open or partial shade. A range of temperature from 10 to 38° C appears to be well suited to the plant. As for the rainfall, the plant is reported to grow naturally in areas with an annual precipitation of 250 cm but it grows well where it is even 500 cm or more if proper drainage exists. It also grows well in areas exhibiting wide disparity in rainfall. Although the plant has been cultivated in all types of climatic regions, tropical areas with sufficient rainfall are the best suited for the plant.

Soil

R. serpentina grows on a wide variety of soils, ranging from sandy-alluvial loam to re-lateritic loam or even hard dark loam. In the natural habitat the plant shows it definite preference for clay or clayey loam containing humus and other organic matter that ensure uniform moisture and proper drainage. The acidic and natural soils are better and the plant does not grow well in saline or alkali soils.

Cultivation

The land to be prepared is initially deep ploughed once in the month of May. After pre-monsoon showers farmyard manure or compost is used at the rate of 30-37 cartloads per hectare. At this stage second ploughing is done followed by two cross harrowing to break the clods. Land is finally dressed by planting.

Use of organic manures, leaf mould and compost has been recommended to increase the quantity of the nutrients in the soil. It was found fruitful to apply a basal dose of 30 kg per ha. of K_2O and P_2O_5 and 20 kg on Nitrogen per hectare before planting. Two top dressing doses on N at 20 kg per ha should also be applied during the growing season every year.

Propagation

The crop can be raised both by seed and by vegetative propagation.

By Seed

As direct sowing in the field does not give good results, seedlings are raised in the nursery before transplanting in the field. The seeds are known to have a very poor and variable germination percentage. About 7-8 kg of good, fresh viable seeds are sufficient to raise a plantation of one hectare.

Before planting the seeds should be soaked in five percent sodium chloride solution. The seeds floating on the surface should be discarded and those, which sink, should be selected. The pre treatment of seeds with some fungicides such as ceresin and captain before sowing not only improves crop stand but checks pre and post emergence damping off. Before sowing the seeds are soaked in water overnight.

(A) Nursery Preparation

The raised beds are preferred which should contain one third quantity of well-rotten farmyard manure and two third of fine soil along with 10 % BHC dust at the rate of 20 Kg per hectare. The seeds should be dibbled about 0.6 cm deep in lines 7.5 cm apart. Thick sowing should be avoided. It is estimated that about 5.5 Kg of seeds sown in 0.05 hectare gives adequate number of seedlings to plant one hectare.

(B) Transplantation

The seedlings are transplanted during the monsoon when general atmospheric conditions are more congenial for their establishment. About 15 cm deep furrow should be dug at a distance of 45 cm transplanting the seedlings and keeping 30 cm spacing between the plants. Seedlings, which have attained a height of 7.5-12.0 cm, are carefully picked up from nursery beds to site of transplantation. A hole, 15-20 cm deep, is made with a pointed stick and seedling is placed in it, and soil is then gently but firmly pressed. The fields are irrigated soon after, planting.

By Vegetative Propagation

When seeds are in short supply vegetative propagation may be practice.

(A) Root

The large taproots as well as lateral secondary rootlets are used for preparing the cuttings of 2.5-5.0 cm long and about 12.5 mm in diameter. Planting is done horizontally in holes 5 cm deep at the advent of monsoon when about 70% them root. It is estimated that about 100 kg root cuttings are needed to plant one hectare. The plants thus raised have high percentages of success but have lower alkaloid content in the roots compared to those raised through seeds.

(B) Root stumps

Approximately 5 cm of root with a portion of stem above the collar is planted in May-June in irrigated fields. Though this method ensures 90-95% or more success, it has a serious limitation as only one plant can be raised from a single stump.

(C) Stem

Stem cuttings, 15-23.0 cm long, with three internodes are considered ideal. Cuttings planted during the early monsoon and kept moist gave 40-46 % success. Treatment of cuttings for 12 hours with B-indol acetic acid (30ppm) initiates rooting within a period of 15 days.

Irrigation Interculture

Usually the crop is irrigated at an interval of 15 days in summer and 25-30 days in winter. For proper development of roots two weeding are considered necessary in the first year, once during the monsoon and the other about the close of the growing season. Only one weeding may be done during the monsoon in the second year. Hoeing is done once manually at the beginning of the growing season during February to March, to maintain satisfactory development of the roots.

Diseases

Rauvolfia serpentina is prone to infection by different groups of fungi and also pests, which causes considerable damage to the crop.

a. Leaf spot

Cercospora rauvolfiae causes small dark brown spots on the upper surface of the leaf and yellowish brown on the lower surface. These spots subsequently merge and cause yellowing and drying of the affected leaves resulting in defoliation. Control can be done by spraying Dithane Z-78 or Dithane M-45. *Mycosphaerella rauvolfia* also causes grayish black small spots, which are oval or round in shape. The lowest leaves are attacked first and the disease progresses upwards.

b. Target leaf spot

Corynspora cassiicola causes dark brown spots on the upper and yellowish brown spots on the lower surface of the leaves throughout the growing season. These spots gradually enlarge into circular spots, 2-20 mm in diameter with concentric zones due to which diseases is named as "target spot". Heavily infected leaves turn yellow and are shed prematurely.

Captan sprayed as 0.25% solution in water, in early June before monsoon and repeated at monthly intervals until November is most effective in controlling the disease.

c. Leaf blotch

This is caused by *cercospora serpitinae* purple, mud coloured botches appear on the ventral surface of the leaf. Disease occurs as a small dot increasing to a circular patch, which later coalesce to an irregular form. In weeks time the affected portion become maroon in colour and the whole leaf gradually dies. Spray of captan and dithane Z-78 or M-45 may prove useful in controlling this disease.

d. Leaf blight and bud rot

Alternaria tenuis manifests as small brownish, dark coloured circular spots with yellowish margin on the ventral surface of the leaf. These spots then enlarge to conspicuous dark brown circular lesion due to which the affected leaves turn brown and subsequently drop. The attack is more common in one-year old plants. Young floral buds and even the fruits may be affected.

e. Anthracnose

Colletotrichum gleosporioides causes the disease and appears as numerous tiny spots of the acrevuli scattered all over but mainly confied to the upper leaf surface. Infected spots enlarge into circular patches and invade the surrounding tissues. Several such patches coalesce and cause drying of the lamina resulting in defoliation. Every infected spot is an acervulur packed with spores.

Cutting and burning the infected shoots, avoiding overhead watering which causes easy dissemination of spores by water splashes, and using dithiocarbonates such as fermate, Dithane and terlate in place of Bordeaux mixture, have been recommended for controlling the diseases.

Pests

Several pests are also reported on *R. serpentina,* which causes extensive damage to the plant.

a. *Glyphodes vertumnalis* (Pyralid caterpillar)

The caterpillars roll the leaves, harbour in the rolls and feed on green matter causing extensive damage to the foliage.

b. *Deliphila nerii* (sphingid moth)

The young leaves are attacked by the caterpillar which feed on them causing defoliation of the plant. The incidence of pest is severe during winter, Dusting 5% dipterex or 2% folidol dust gives effective control.

c. *Scarbacide* sp. (Grub)

The pest profusely attacks the main roots of the seedling during June resulting in their drying up. Mixing BHC (10%) with the soil at the time of preparation of land can control the grubs to some extent.

The other pest's reprted to cause damage to the crop are *Deaphania nilgirica* (moth), *anomala polita* (Cockchafer grub), and *Saissetia niga* (black bug).

Harvesting

Under irrigated condition, the ideal age for harvesting of plants to yield exploitable size of roots is from 2-3 years after planting. Roots should be dug in winter when the plant is completely defoliated. A light irrigation should be given advance to facilitate easy digging of roots. The marketable root goes 40-60 cm deep and may have to dug out by subsoiler or mould board plough. All the secondary and tertiary roots should be dug out and collected because they are rich in alkaloids. The roots should be cleaned off the adhering soil and washed if necessary. The root bark, which constitutes 40-60% of the whole root, is richer in alkaloids.

Drying and storage

Cleaned roots are kept for air-drying on dust free pucca floor till they completely dry and become brittle. These can be stored for long by packing in gunny bags. If roots are to be stored for longer period they should be sun dried or exposed to sun periodically to maintain quality of the roots. To channelize the food material for better root growth, fresh flower tops designing to develop when the crop is one year old should be clipped and removed.

Seeds should be collected only from the flowering flush appearing during September-November (14-16 months after the crop growth). The matured berries, dark purple in colour, should be collected and kept overnight. These should be cleaned the next morning by rubbing on a rough surface or on gunny bags and washed in water.

Seed should be air dried for 10-15 days and stores in air right containers. The freshly harvested seed gives better germination and maintains viability upto 6-8 months.

Yield

The plants raised from seeds give the maximum yield of roots varying widely from 0.1 to 4.0 kg per plant. It is estimated that with a spacement of 60 x 30 cm and survival of 80%, the yield of roots per ha in case of plants raised from seeds is about 1,175 kg (air dry basis) as compared to 175 kg in plants raised from stem cuttings and 345 kg in plants raised through root cuttings. An irrigated 2-year-old plantation on sandy-clay-loam soil is reported to yield 2,200 kg and a 3-year-old plantation 3,300 kg air-dried root per hectare.

Chemical composition

The root owes its pharmacological activity to the presence of several alkaloids of which reserpine is the most important. Others are rescinnamamine, resepinine, serpentive, aserpentinine, sepogine, yohimbine, 3-epi-d-yohimbine, ajmaline, ajmalinne, ajmalcine, iso-ajmalcine, chadrine, rauvolfinine renoxidine, reserpiline and tetraphyllicine.

The total alkaloidal content of the roots ranges normally form 1.7 to 3.0 %, which are mostly concentrated in the bark (about 90%). Smaller quantities are also present in leaves, stems and seeds.

Besides the alkaloids root contains oleresin and a sterol, serposterol, the rasin fraction is physiologically active. In addition to *R. serpentine* it is also found in *R. canescens, R. nicrantha* and *R. vonittoria*. Rescinnamine is also used as a hypotensive agent and tranquilliser. It is supposed to be relatively free from side effects.

GUGGAL
[*Commiphora wightii* (Arn.) Bhan.]

GYANENDRA TIWARI
Dept. of Plant Physiology, JNKVV, Jabalpur (M.P.)

Botanical Name : *Commiphera wightii*
Family : Burseraceae

Vernacular names

San.-Gugguluh, Mahisaksah; Hin.-Gugal, Guggul; Kan. -Guggulu; Mal.-Gulggulu, Mahisaksagulgulu; Mar.-Gugglu; Tam.-Gukkulu, Mahisaksi, Kukkil; Tel.-Guggulu, Maisakshi; Eng.-Indian bedellium tree; Trade-Guggal.

Introduction

This plant is a source of Indian bedellium, a oleo-gum-resin obtained by incision of the bark. The guggul resin is a complex mixture of steroids, occurs in vascular or stalactitic pieces, pale-yellow, brown or dull-green in colour with a bitter aromatic taste and balsamic odour.

Botanical description

A small tree or shrub up to 3-4 m high, the branches are crooked, knotty, aromatic and end in sharp spines. The bark is papery and peels in strips from the older parts of the stem. The leaves are sessile, alternate or fascicled, 1-3 foliate, the leaflets are glabrous. The flowers appear in groups of 2 or 3 usually red and rarely pinkish-white in colour. Fruit ovoid drupe, red when ripe.

Geographical distribution

Its origin in Africa and Asia and is widely distributed in the tropical regions of Africa, Madagascar, Asia,

Australia, Pacific Islands, India, Bangladesh and Pakistan. But the plants are found growing naturally in arid zones of Western Indian. It is distributed in the states of Rajasthan, Tamil Nadu, Assam, Gujarat, Maharashtra and Karnataka.

Medicinal uses

The resin is largely used as incense and as a fixative in perfumery and in medicine. The oleoresin causes an increase of leucocytes in the blood and stimulates phago-cytosis. It acts as a diaphoretic, expectorant and diuretic and is said to be a uterine stimulant and an emmenagogue as well. It is highly effective in the treatment of obesity, arthritis, indolent ulcers, weak and spongy gums, pyorrhea, alveolaris, chronic tonsillitis and pharangytis, ulcerated throat and chronic dyspepsia. Inhalation of the fumes of burnt guggul is recommended in hay fever, acute and chronic catarrh, chronic laryngitis, chronic bronchitis and phthisis. It is an ingredient of ointments for ulcers.

Chemical constituents

The oleo-gum-resin is mixtures of resin (about 61%), gum (about 29.3%), volatile oil (about 1.45%) and occasionally combined with other substances, mainly myrcene, dimyrecene and polymyrecene.

Soil and climate

It grows well in sandy to silt loam soils, poor in organic matter but rich in several other minerals with moisture-retaining capacity. It prefers a warm, dry climate for a good yield of oleo-gum-resin.

Improved varieties

'Marusudha'

Agro-technique

Land preparation

The land is prepared well in advance of the rainy season by 2-3 ploughings and laid into plots of convenient sizes.

Pits of the size 0.5x0.5 m^3, are dug at a spacing of 3 to 4 m, in rows. The pits are filled with FYM and topsoil and mixed with Aldrin (5%), to prevent any damage to the plants due to termite attack.

Mode of propagation

It can be propagated by seeds or vegetatively viz. stem-cuttings, air layering, etc.

By seeds

This is not a common method of propagating this plant, because the seed germination is slow and is also very poor (5%) due to the presence of the hard seed coat. This achieves a good rate of germination; the seeds are mechanically scarified with sand paper and are kept under running water for 24 hours. The seedlings may be raised in polythene bags during the kharif season and, after hardening, may be planted in the main field.

By stem-cuttings

About 15-20 cm long and 10 mm thick, semi-hardwood cuttings are taken and the cut end is treated with IBA or NAA growth-regulator solutions and planted in well-prepared and manured nursery beds during the months of June and July. The beds are irrigated lightly after planting and regularly thereafter. The cuttings sprout in 10-15 days, they grow well and are ready for planting in the main field after 10 to 12 months, during the next rainy season. The percentage of rooting in stem cutting is around 80 to 94%.

Transplanting

The rooted cuttings are planted in the pits during the rainy season. As the plants grown, they are trained properly by cutting the side branches.

Manure and fertilizers

The crop has not shown good response to fertilizers, except to low-level or nitrogen fertilization. Hence, urea or ammonium sulphate at 25-50g/bush is given twice a year before Irrigation.

Irrigation

Providing light irrigation during the summer season supports good growth of the plant.

Interculture

It is confined to one weeding and hoeing in the early stages of growth. But the soil has to be stirred around the bushed twice in a year. This practice of soil stirring is found to increase the growth of plants.

Pruning

The guggal stems purned in May contain the maximum amount of guggal sterons (0.06%).

Pests and diseases

Guggal plants are attacked by a leaf-eating caterpillar, white fly, which sucks the sap of leaves as a result of which the leaves turn yellow and finally drop off. The rooted cuttings also suffer from termite attacks in the early stages of growth. The crop is mainly suffering from leaf spot and bacterial leaf-blight diseases.

Harvesting (gum-tapping)

The plants attain normal height and girth after 8 to 10 years of growth, when they are ready for tapping. For tapping the gum, which is present in the balsam canals in the phloem, bark-deep, i.e. a shallow, incision is made on the bark. If the incision is too deep, the plant either dies or yields little resin in the following years.

Production and yield

The yield of gum resin from each plant is about 700-900g. The yield in the initial year will be very low, but it gradually increases in the subsequently years. After 5 years, from a well-managed field, we may get about 2.0-quintals/ ha gum resin.

Sadabahar (*Periwinkle*)

Dr. A.S. Gontia
Principal Scientist,
Deptt. of Plant Physiology, JNKVV, Jabalpur (M.P.)

Botanical Name	:	*Catharanthus roseus (L.) G. Dass*
Family	:	Apocynaceae
Common name	:	Sadabahar

Introduction and Economic Importance

Sadabahar belongs to the family Apocynaceae of Botanical kingdom. Its botanical name is Catharanthus roseus (L.) G. Don. also called Baramasi or Sadasuhagan in Hindi, Billaganneru in Telgu and kasi kanigale in kannada. It has been recognized as one of the most important plants for medicinal use.

Sadabahar is recorded as back as 50 B.C. in folk-medicine literature of Europe as diuretic, antidysenteric, haemorrhagic and wound healing. It was known for use in the treatment of diabetes in Jamaica and India. The leaves are source of alkaloids, vinblastine and vincristine, which are recognized as anticancer drugs. Raubasine also known as ajmalicine and reserpine present in the roots are recognized as anti fibrillic and hypotensive agents. The alkaloid vincaleukoblastine (VLB) obtained from the plant has been used in the Hodgkin's disease, a cancer that attacks lymph glands, spleen, liver and choriocarcinoma.

USA is the world's largest user of raw material. Hungary has been one of the major consumers of the leaves. West Germany, Italy, Netherlands and UK are interested in roots for extraction of raubasine.

Catharanthus is a tropical genus idigenos to Malagasy, spreading to Mozambique, India, Philippines, Australia,

South Vietnam, Sri Lanka etc. There are three variants in C. roses (i) Rose-Purple flowered (ii) white flowered and (iii) white flowered with rose-purple spot in the center; the first type is being cultivated because of its higher alkaloid content. Recently two white flowered varieties Nirmal and Dhawal have been released by CIMAP Lucknow, which, although equal in active principles are reported to yield a higher biomass.

Description of Plant

It is a perennial herb, up to 0.9 m in height. Stem erect and branched, leaf simple, opposite and petiolate, Inflorescence-racemose, flowers in axillary pairs, calyx and corolla-5, androecium-5 stamens, filaments epipetalous gynoecium carpels 2 free below but united in the astylar region, ovaries 2 free single style with dumbel shaped stigma, seeds small, black in colour 1 g contain about 835 seeds.

Soil and Climate

The cosmopolitan distribution of the plant shows that it has no specific soil and climate requirements. However, the growth is much better in tropical and subtropical areas. In India it is cultivated in a large scale in Tamilnadu, Karnataka, Andhra Pradesh, Gujrat, Madhya Pradesh and Assam. The plant also grows in sub-tropical areas of northern India, but the growth is slow due to extremely low temperature during the winter.

In can be grown on any type of soil except those, which are highly saline alkaline or water logged. Light soils, which are rich in humus are most suited for its cultivation. A well-distributed rainfall of 100 cm or more is ideal if the plant is to be grown as a rainfed crop.

PROPAGATION

It is mostly propagated from seeds. Seeds can be sown either in a nursery and then transplanted or sown directly in the field.

1. Direct sowing can be adopted during the mansoon months, particularly if large area has to be cultivated. About 2.5 Kg of seeds are required per hectare. Seeds are mixed with 25 Kg of moist and fine sand to ensure even distribution. They are sown in rows 45 cm apart and subsequently seedlings are thinned out in rows maintaining distance of about 30 cm between the plants.
2. Transplanting - If seeds are scare, transplanting can be adopted with advantage since only about 500 g of seeds will be enough to plant one hectare. In this case, the seeds are sown in nursery buds, two months in advance, so that the transplanting coincides with the onset of mansoon, an area of about 200sqm. under nursery gives enough seedlings for transplanting one-hectare land. The seeds take about 10 days to germinate and about 60 days to reach transplanting stage. Transplanting is done at 45 X 30 Spacing. One hectare requires about 74000 seedlings.

Propagation Through Cuttings

Propagation can also be carried out through cuttings. The cuttings of 10-12 cm having three to four nodes with axillary buds would be more suitable for this purpose. Cuttings can even be placed in trays after dipping in solution of 100 ppm IAA before planting in field. After about one month with the emergence of 4-5 leaves in cuttings the cuttings are ready for transplantation.

Manures and Fertilizers

Prior to sowing farmyard manure be applied @10 ton/ha. A basal does of 50 kg/ha. N 50 kg/ha P_2O_5 and 50 kg/ha K_2O are also incorporated in the soil. The remaining nitrogen be applied @ 100 kg/ha in two split doses. First application is made 10-15 days after transplanting and second one month later.

Weeding

The crop requires two weedings in the initial stage. First weeding is done about 60 days after sowing/transplanting and second 60 days later.

Irrigation

Places where rainfall is evenly distributed throughout the year, the plants do not require any irrigation. However, the areas where the mansoon is restricted to a particular period, 4-5 irrigations once in fifteen days during February, March and April months are needed to get optimum yield.

Harvesting

Roots - The crop is harvested about 12 months from sowing. The crop is cut at about 7.5 cm. above the ground and dried in shade. For collection of roots the field is irrigated and after some time it is ploughed and roots are collected. The roots are thoroughly washed and dried in shade.

Leaves, Stems and Seeds - It there is demand for leaves, two leaf stripping at 6th month and 9 month can be carried out. The Third stripping of leaves is obtained when the whole plant is harvested. After harvesting, the whole plant is dried in shade. At this stage, light threshing will separate the seeds. The leaves and stems are also collected separately. Seeds have the tendency to mature throughout the year hence soon after maturity seeds should be collected to avoid shattering.

Yield - The production of leaves, roots and seeds was found to be around 3-6 tons, 1.5 - 3.0 tons/ha. and 4.5 tons / ha, respectively.

Pests and Diseases

The plant is very hardy and resistant to the attack of fungi and insects, Occasionally, some plants have been found to suffer from little leaf disease which is caused by phytoplasma (micoplasma) resulting is stunted growth of the plant. The disease can be checked by uprooting and destroying the affected plants. Since the jassids are the vectors of the disease they are required to be controlled by spraying Nuvacron trizophos or Metacid @ 2ml/litre. Recently die back, caused by Pythium aphanidermatum, Phytophthora spp and Rhizoctonia has been identified to damage the crop. In this disease drying and blackining starts from shoot tip basewards which results in drying of whole plant. The disease can be controlled by spraying the Bordeaux mixture which is the combination of copper sulphate, lime and water in the ratio of 5:5:50. Spraying done at an intervals of 10 days thrice was found to be most effective. Mulching between the rows at seedling stage with any straw like dry spent e.g. Citronella or palmers reduces the incidence of die-deck to a considerable degree. One die back resistant strain WH 10 of Sadabahar has been developed at CIMAP Regional Center, Bangalore.

Active Constituents and Uses

At present more than 100 alkaloids have been isolated from the sadabahar. The percentages of total alkaloids in deferent plant parts are as follows: Leaf (0.6 –0.67), Stem (0.26-0.31), Flower (0.005 - 0.007), Root (0.78 - 1.22), root back (4.5-9.00). Apart from vinblastine and vincristine, some of the other important alkaloids present are reserpine, ajmalicine (raubasine) and serpentine Vinblstine in the form of vinblastine sulphate is available in American market under the trade name 'VELBE' and vincristine sulphate as ONCOVIN. The former is mainly used in the treatment of Hogdkins disease and other lymphomas and choriocarcinomas. Vincristine sulphate is more effective against acute leukemia particularly in children.

SENNA (*Cassia angustifolia* Vahl.)

AASHUTOSH SHARMA
Research Associate
Deptt. of Plant Physiology, JNKVV, Jabalpur

Family : Caesalpinoidae (Leguminosae)

Engilish name : Senna

Indian name : Svarnapatri (Sanskrit)
Sanay, Sona ka patt (Hindi)
Nelavakai (Tamil)

Introduction

Senna is a small dry land perennial shrub, also called as sonamukhi or sennai, which is botanically called as *Cassia angustifolia* vahl is a native of Yemen, South Arabia. It has come to India from Arab. In their native land, these species grow on arid tracts as perennial bushes. However, senna is maintained as annual herb when cultivated (duration 5 to 7 months). Tamilnadu (Tirunelveli, Madurai, Tiruchirapali districts etc.) has been the principal state in the production of the plant (8000 ha to 10,000 ha area) both under rainfed and irrigated conditions. Recently Rajasthan has taken lead in the cultivation of senna (Barmer, Jaisalmer, Bikaner, Nagaur, Pali, Churu, Sikar, Jalore and Ganganagar). Although most plant parts contain sennosides (glycosides) but leaves and pod shells contain highest concentration of sennosides (glycosides) A, B, C and D. Pods contain higher percentage of total sennosides (3-5 percent) then of foliage (2.5-4 percent). Demand for leaves is higher for use in herbal tea, bakery products and other home preparations in west and central Europe. The UK and USA have preference for pods. A major part of the produce is exported in the form of leaves, pods and sennoside concentrates, though several pharmaceutical houses utilize it for the manufacture of calcium sennoside granules, tablets and syrups for marketing within the country as well as outside the country. Indian pharmaceutical industry uses about 100 tons of leaves and pods. The total world requirement is about 10,000 tons of leaves and pods. India is the major exporter and exports upto 5000 tons worth Rs. 20 crore every year.

Botanical description

Cassia genus comprises about 500 species, of which 20 are found in the native flora of India. There are two sources of the senna drug, namely, *Cassia angustifolia* vahl (native of South Arabia, West Asia) and *C. acutifolia*. Del. (Sudan, East Africa) are exotic to India. Indian produce comes entirely from the cultivated crop of *C. angustifolia*. It is a small perennial undershrub, below 1m in height with ascending branches. The leaves are large, compound and pinnate. The full-grown leaflets are bluish-green to pale-green in colour and emit a characteristic fetid smell when crushed. The flowers are bright yellow in colour, arranged in axillary, (on subterminal) erect, many-flowered racemes. The flowers are not season-bound and are borne between 35 to 70 days of age, depending upon the time of sowing, and its season of cultivation. The pods appear immediately after flowering, are slightly curved, 3.5 to 6.5 cm long and up to 1-5 cm broad, green in the beginning changing to greenish-brown to dark brown on maturity and drying. Each pod has 5 to 7 ovate, compressed, smooth, dark-brown seeds.

Soil

This crop prefers well-drained sandy loams to laterite soils. However, it can grow in marginal and submarginal lands with pH 7 to 8.5. This crop is highly sensitive to water logging and heavy rains destroys the crops.

Climate

Senna is a rainy season crop and requires low rainfall during the whole life span. It is also a sun-loving crop and requires bright sunshine for its successful growth. Heavy rains and cloudy weather during growth are harmful to the crop. An average rainfall of 25-40 cm. distributed from June to October is sufficient to yield good harvests.

Variety

"ALFT-2" by Anand Gujarat Agriculture University, released late flowering type through selection, which remains in a vegetative stage till 100 days and is suitable for growing exclusively as a leaf crop. The CIMAP Lunknow has also released a high-yielding variety named 'Sona'.

Sowing

The crop is raised from seeds. Senna can be grown as an early summer (Feb-March) or a winter (Oct.-Nov.) crop. In a well prepared land seeds are sown in lines 30-40 cm apart at 1.5-2.0 cm depth during Feb-March in Northern India and Oct.-Nov. in Southern. Soil moisture should be enough for germination even if it is grown in dry land fields. Delay in sowing reduces the vegetative phase and ultimately affects the yield of the drug. Plant population is the major factor for obtaining desired level of production. Therefore, seed treatment with fungicide, optimum time of sowing, and depth of the seed at the time of sowing and soil moisture is the important factors to be taken care off.

Since the seeds have a hard and tough seed coat, a certain amount of abrading of its surface is necessary to induce quick germination. Soaking the seeds for 10-12 hours before sowing is reported not only to give 100% germination, but also the stand of the plants grown is even. Germination is complete within a week. The crop may be thinned after 30-35 days to maintain plant-to-plant distance of 30 cm. Seed rate of 12-15 to 20 Kg/ha with 70-80% germination is sufficient to have a population of 70-75 thousand plants/ha, a desired level for economic yield. It is found that treating the seeds with Thiram, captan or Agroson G.N. at 2.5 g/Kg protects the seedlings from damping-off and seedling blight disease, which are common occurrence.

Manures and fertilizers

General recommendation of fertilizer for obtaining economic senna yield is 80:40:20 Kg/ha NPK. Nitrogen application should be splitted in doses each after thinning, 1st and second picking.

Irrigation

One-irrigation 7 days after sowing is very important if rains not commence. Senna can be economically grown under rain fed conditions. The crop needs no irrigations except during prolonged drought. However, when it is grown as a semi-irrigated crop, the yield increases considerably. In all, about 5-8 light irrigations are enough to raise a good crop of senna as heavy irrigations are injurious to crop. Irrigation after each picking and top-dresses with Nitrogen increase the yields.

Pests and diseases

Pests

Several leaf-eating caterpillars larvae feed on the green senna leaves. Spraying of Carbaryl (4g/l) periodically in the growing season controls the infestation. Pod borer also reported to attack the pods and can be controlled by spraying Endosulphan (0.05%) or Carbaryl (0.25%) at an interval of 10-15 days.

Diseases

Leaf-spot disease is the most serious disease and causes severe damage to the crop. The infested leaves drop off, resulting in poor yield. The sennoside content in the leaves is inversely proportional to the intensity of the disease. The spraying of 0.15% Dithane M-45 at fortnightly intervals 3 times in a period of 5-6 weeks has been found to control the disease effectively. The disease is caused by *Alternaria alternata* (Fr.) Keisel., *Phyllosticta* spp and *Cercospora* spp.

Damping off of seedlings is the most devastating disease, which is caused by *Rhizoctonia bataticola*. The disease spreads rapidly if the fields are affected by stagnating moisture at this stage. Treatment with thiram or captan at 2.5 g/kg of seeds protects the growing seedlings. At a later stage its other physiological form called *Macrophomina phaseoli* develops, it causes dry rot in the crop and kills the plant. The fields can be drenched with 0.2% Brassicol or 0.5 to 01% Rhizoctol, but this only gives partial control.

The leaf-blight disease is caused by *phyllosticta* spp. Usually occurring at the later stages of growth, in September-October. Cloudy days and humid weather conditions favour the spread of the disease. The spraying of 0.15% Dithane M-45 at fortnightly intervals can control the disease.

Weeding

Two weeding at 30 and 60 days after sowing are essential to control the weeds.

Rotation and Intercrop

In Tamil Nadu, senna is grown after the paddy. Senna-mustard or Senna-coriander rotations are found better under AICRP on medicinal plants. The Senna crop can be grown as inter crop in between plantation crops. In medico forestry ecosystem it should be grown with Aonla, *Oroxyllum indicum* or *Gmelina arborea* etc. under fallow and marginal lands.

Harvesting

It has been found that young senna leaves and pods contain a high sennoside content but since the produce is sold on the basis of weight, a balance between weight and content has to be made to choose the right stage for harvest. Picking of the upper branches is done by sickles when the lower leaves become yellow, brown at an interval of 50,90,130 days in Delhi condition and 90, 110 and 130 days at Anand are recommended. The harvesting should be done during the clear days to avoid rotting of leaves. The harvested branches are kept in field in small bundles for drying and it should be turned on next day. Later on (3rd day) it should be collected on the threshing floor where it should be dried in shade to avoid browning of leaves.

Yield

Well-managed irrigated crop yields about 15-20 q dried leaves and 7-10 q/ha dry pods. Under rainfed 10 q/ha dried leaves and 4 to 5 q/ha dried pods yields are reported.

Economics

The cost of production varies 15 to 20 thousand with net profit ranging from 10 to 20 thousand/ha depending upon the prevailing rates.

Chemical constituents

Leaves: glucose, fructose, sucrose, pinnitol and sennoside-C, Mucilage: galactose, arabinose, rhamnose and galacturonic acid. Pod- Sennoside A and B, glycosides of anthraquinones, rhein and chrysophanic acid, Seed: B-Sitosterol.

Uses

Leaf and pod- laxative especially useful in habitual constipation. Safe purgative for expelling intestinal worms.

1. Constipation

Senna leaves are safe and effective purgative and can be used by children and old person. Senna pods can also be used as a purgative but slower than leaves.

2. Intestinal worms

Senna can be used for killing the intestinal worms.

3. Skin disorder

A paste of the dried leaves made with vinegar as a base can be used for acne, eczema, and pimples.

4. Preparation

Its leaves can be given as infusion, decoction, powder or confection. One of the best preparations is made by infusing leaves, 60 g with ginger 4 g in one liter of water in a covered vessel for 15 minutes.

5. Dose

0.6 to 2g of leaves or pods soaked in 125 ml of warm water for about 12 hours as infusion whereas, it is usually 30 mg of sennosides per tablet.

GILOY (*Tinospora cordifolia* (willd) miers ex H.K.F. Thoms.)

S.K. DWIVEDI
Associate Professor
Deptt. of Plant Physiology, JNKVV, Jabalpur (M.P.)

Botanical Name	:	*Tinospora Cordifolia*
Family	:	Menispermaceae

Vernacular names

Sanskrit - Amrita, Guduchi, Hindi - Giloy, Giloe, Gulancha, Bangali - Giloe, Gulanch, Marathi - Gulwel, Gujrati - Galo, Oriya - Gulochi, Panjabi - Batindu, Malayalm - Amrita, Chittamrita, Chittamritam, Tamil - Amrutavalli, Kannada - Amritaballi, Telgu - Tivantiva, Tippatige, Guduchi, English - Gulancha, Tinospora.

Botanical description

Giloy belongs to the family menispermaceae. It is a big, perennial deciduous, flesh twiner which climbs on trees like neem, mango etc. Many roots arise from stem and remain pendent. The bark is dusty brown or pale yellowish white and papery on its removal inner green flesh is seen. Leaves are simple; alternate, cordate with entire margins, glabrous and 5-10 cm broad. Flowering occurs in December-January. Flowers are small yellow in lax racemes, male flowers in clusture and female flowers usually solitary. Fruits are pea sized, oval drupe, shining red when ripe.

Distribution

The plant is widely seen in tropical countries like India, Myanmar, Sri Lanka. In India it occurs in forests up to 1000 to 1200 m elevation. It is mostly found trailing on forest trees, fences and hedges along the boundries.

Climate

Giloy requires a warm humid tropical climate.

Soil and soil preparation

It can be grown in all types of soils from wasteland to sandy loam soils rich in organic matter and having well drainage system.

In the month of May or June pits of size 30 cm^3 are to be taken, filled with 10 Kg FYM and covered with top soil and also with a small quantity of B.H.C. powder.

Propagation and planting

It is propagated vegetatively by stem cuttines. About 10-15 cm long stem cuttings having at least 2 nodes are planted in the field or polybags. Treatments of cut ends with hormones gives better results. The usual planting time is with the onset of monsoon in May-June. Usually it is planted along bondaries or nearby tall trees and allowed to trail on the trees or hedges and hence regular spacing is not followed but a spacing of 5x5 m gives better results.

Manures and fertilizers

Though it makes satisfactory progress with little manuring and fertilization. Normally a fertilizer dose of 20g urea and 40 g super phosphate per plant ensures a vigorous plant after fertilizer application the climber should be irrigated.

Interculture

One weeding is required one month after planting. Later on the weeding may be done if required.

Irrigation

Giloy has a less water requirement and it should be irrigated frequently, once in 10-15 days during the dry period and ones during winter season.

Pest and diseases

The plant is sufficiently hardy and free from pests however, proper control measures may be taken if the crop is affected by insects and diseases.

Harvesting and storage

The stem is most economic plant part. Harvesting can be commenced after one year and usually partial harvesting is followed depending on the necessity. The stem and root should be collected in hot season when the concentration of bitter principle is the highest. A full-grown well-ramified plant may give 2-3 kg vines, which are cut in to small pieces and traded either fresh or dried.

Production and yield

The stem is cut in to pieces of 8-10 cm and sold in the market, after 5 years of planting 20 kg fresh stem is produced per climber. An average of 40-quintal fresh and 8-10 quintal dry stem may be obtained per hectare.

Chemical constitution

The stem contains cordifol, tinosporidine, a diterpenoid-tinosporide and perberilin other constituent of stems are hepta cosanol and b-sitosterol. Leaves contains tinosporon, tinosporic acid; tinosporol, d-sitosterol and cordifolide, afuranoid bitter principle - tinosporine and quaternary alkaloids - magnoflorine and tembetariene.

Medicinal uses

The stem is used in medicine. It is acrid, sour, sweet, thermogenic and astringent. It is useful in vitiated conditions of *vata*, burning sensation, body pains, intestional worms, inflammations, intermittent fevers, chronic fevers, vomitting, indigestion, flatulence, loss of appetite, gout, cardiac debility, skin diseases, anemia, cough, asthma, general *debility*, seminal weakness, uropath and splenopathy.

The Giloy is well known Indian bitter and is prescribed in fevers and as an antidiabetic. The starch obtained from the plant is considered easily digestible and highly nutritive and given to the patient with poor digestion. The plants growing on neem tree (*Azadiracta indica*) are considered more efficacious; as it is said to be incorporate the medicinal virtues of neem also.

In Ayurveda, the drug is gives with lendipiper (*Piper longum*) root and ginger for improving digestion. In migraine the juice of the plant with honey is administered. In vaginal inflammation, a decoction of giloy, triphala and danti or wild caster (*Balliospermum montanum*) is recommended.

In Unani system mostly 'Sat Giloe' is incorporated in the preparation 'Arq Giloe' prepared from the fresh plant is considered a febrifuge while "Arq maul" laham mako - kashiwala is a general tonic. The dose of giloy decoction is 40-80 ml and powder is 2-4 g. The giloy is a popular remedy for snakebite and leprosy. As a tonic it is best given in infusion with or without milk. It is a constituent of several preparation like Guduchyadichurna, Gududyadi kwath, Guduchilouha, Amritarista, Sanjivani vati, Guduchi taila, Amritastak kwath etc, Several oils for external applications are prepared with amrita and applied to skin diseases, rhumetic alterations and other nervous complaints. A small quantity of bruised stem soaked for three hours in half a litre of water and strained combined with ammonium acetate is administered in intermittent and milder forms of fevers.

GURMAR (*Gymnema sylvestre* (Retz.) R. Br. Ex. Schulters)

S.K. DWIVEDI
Associate Professor
Deptt. of Plant Physiology, JNKVV, Jabalpur (M.P.)

Botanical Name : *Gmnema sylvestre*

Family : Asclepiadaceae

Vernacular names

Sanskrit - Merasringi, madhunasini, Hindi - Gurmar, merasringi, Bangali - Merasingi, Malayalam - Chakkara kolli, Madhunasini, Tamil - Sirukurumkay, Adigam, Sakkaraikkolli, Telgu - Podapatra, Marathi - Kavali, Kali Kardori, Vakundi.

Botanical description

Gurmar *(Gymnema sylvestre (Retz.) R.Br. EX Schulter)* belonging to the family *Asclepiadaceae*, is a large, more or less pubescent, woody climber. Leaves 2.25-5 cm long and 1.2-3 cm broad but occasionally may go to 10-12.5 cm in length, simple opposite, elliptic ovate or ovate- lanceolate to ovovate upper surface dark green, shining, under surface pale green. Shortly pubescent at venation. Flowers are small, yellow, in umbellate cyme. Follicles lanceolate 7cm long. Fruits are slender two allied species G. *hirsutam* found in Bundel khand, Bihar and western ghats and G. *montanum* growing wild in the Eastern ghats and konkan are also used for the same purpose and are called gurmar.

Distribution

It is a tropical climber and mainly grows in the Western Ghats, konkan, Tamilnadu and some parts of Bihar. The plant is cultivated in the plains of India but drug is mainly imported from Afganistan and Iran.

Climate

It can be grown in all types of climatic conditions.

Soil and soil preparation

Gurmar can be grown in all types of soil preferably with good drainage system. During the month of February-March pits of 30x30x30 cm size are to be taken at the distance of 5x5 m filled with 10 Kg FYM and soil mixture.

Plant propagation and planting

Plant can be propagated both by seeds and stem cuttings. Seedlings are to be raised in polybags. The 3-4 months old seedlings are transplanted in the pits from polybags. Trailing can be facilitated by erecting poles and tying the plants to the poles. The plant will attain good spread within one year.

Manures and fertilizers

Regular application of organic manures is essential for vigorous growth. The 20 g urea and 40 g single super phosphate can be given per plant frequently. An Irrigation should be given just after each fertilizer application.

Interculture

Regular weeding is recommended.

Irrigation

Crop should be irrigated regularly.

Pest and diseases

The plant is not affected by any serious insect pests or diseases.

Harvesting and storage

Leaves can be collected from the first year onwards at an interval of one week. This can be continued for 10-12 year. Generally the leaves are collected in the month of

December and January. The collected leaves are dried in shadow. The roots of gurmar can be uprooted in summer season, which may be thoroughly washed, sun dried and stored.

Yield and production

About 10 quintal dried leaves per hectare may be obtained under favorable circumstances.

Chemical constituents

Leaves gave several o-iso-prophylidene derivatives of gymnemagenin, a hexahydroterpene, crystalline gymnemagenin, gymnestrogenin, and gymnemicacid (the antisweet principle), which was a complex mixture of atleast nineclosely related acidic glycoside. On hydrolysis, various genins including gymnestrogenin and its derivatives were obtained. Leaves also contains hentriacontane, pentatriacontane; a and b-chlorophyll, phytin, antraquinone derivatives inositol and d-quercitol.

Medicinal uses

The leaves when tasted are saltish and acidic and they supress the activity of taste of tongue for sweet taste and for this reason it was considered that it destroys the sugar, hence the name Madhunashini or Gurmar and has been prescribed as an antidiabetic. The dose of powdered leaves is 15-30 g with water. Its effect on the tongue lasts for 3-4 hours. It is not used in unani medicines.

Root has long been reported as a remedy for snakebite. Leaves triturated and mixed with castor oil are applied to swollen glands and enlargement of internal viscera as the lever and spleen. The drug is used to strengthen the function of heart, cure jaundice, piles, urinary calculi, and difficult micturition and intermittent fevers. The drug enters into the composition of preparation like Ayaskriti, Varunadi Kasaya, Varunadighritam, and Mahakalyanakaghritam etc. They suppress the activity of taste of tongue for sweet taste. The crude drug as well as its dried aqueous extract is mainly used in bronchial troubles.

SATAWAR
(*Asparagus racemosus Willd.*)

A.B. TIWARI
Scientist
Deptt. of Plant Physiology, JNKVV, Jabalpur (M.P.)

Botanical Name : *Asparagus racemosus*

Family : Lilliacc

Common Name : Satawar

Introduction

According to wealth of India the genus Asparagus comprised of 5 species viz., *Asparagus adscendens* Roxb., *A. Filicinus* Buch-Ham. Ex. D. Donog *A. gonocladus*, Baker, *A. officnalis* L. and *A. racemosus* Willd. In Bhavprakash Nighanta two types are described viz Chhoti Satawar (*A. racemosus*) and Bari Satawar (*A. sarmentosus*). Racemosus type was considered as Shatavri, Shatpadi, Narangi, Shatwirsh, Satawar or Satmuli and Sarmentosus as Mahashatavri, Shatmuli, Urdhwa kantika, Sahaswirya etc.

Asparagus racemosus is a perennial spenous, climber branches midified in to leaves-cladode 2-6 together 1.3 to 2.5 cm long, flowers white, in 2.5 cm long racemes, roots fassiculated tuberopus, fruits a globose berry green and red on ripening, and seeds black.

It is distributed throughout tropical and sub tropical regions of India. The good quality Satawar comes from Ratlam district of Madhya Pradesh and Nepal.

Climate

This can be cultivated in tropical, sub tropical, temperate climate; Forests on Nepal and Himalaya and mixed forests of Madhya Pradesh are the natural places.

Soil

Well-drained sandy loam to loamy soils are better. However, it can be cultivated in all types of soil having normal pH. Fallow lands and forestlands are also suitable for this crop. It can be cultivated all over India even in undulated unculturable lands.

Seed rate

10-12 kg seed/ha is required.

Sowing period

Seeds are sown in June in nursery in poly bags or in beds. Seeds germinate within 10-12 days. One-month-old seedlings are transplanted. The planting from the discs of old plants can also be done but it will be uneconomical.

Method of planting

During July-August one month old seedlings should be planted at a distance of 50 cm within the rows in 50 cm apart rows under irrigated conditions, while a planting distance of 50 x 25 cm should adopted under unirrigated conditions.

Manures and fertilizers

Good quality FYM @ 20-25 t/ha is applied during the field preparation to obtain the higher root yield.

Irrigation

From October to March it should be irrigated at an interval of one month and during summer in 15-20 days interval. However, it can be cultivated under un-irrigated conditions, but production will be low.

Weeding

Weeds should be removed within a month to reduce the competition. Further weeding should be carried out as per need.

Stacking

To support the climbing stem, the bamboos sticks or tree branches should used for stacking.

Pest and diseases

The crop remains normally free from pests and diseases.

Harvesting

The roots are dug out after two to three year. The roots are cleaned and graded in to smaller and bigger size and cut in to pieces of appropriate size and peeling is done and dried in the sun. Discs with smaller roots should be separated and must be kept for planting. The dried root when become brittle (8-10% moisture stage) must be stored in bags for trading. The fruits are picked during February March when they become red. Fruits are pressed and washed in water to get clean seeds, which are dried in the sun and stored for sowing.

Yield

The yield of fresh root is 300 to 500 q/ha, which remains (30 to 50 q/ha on drying) 10% of fresh roots.

Intercropping

This crop can be grown as inter crop in plantation crop and in Agro-forestry system. It will not require extra support for climbing. *Gmelina arborea, Oroxylum indicum,* Aonla or othor tree species will be suitable.

Chemical composition

Faciculated root contains saponins shatavavins I-IV flowers yield quercetin hyperoside rutin, diosgenin and a flavonoid glycoside.

Use

Antidiarrhoetic, refrigerant, antidysenteric, diuretic, demulcent, galactagogue antispasmodic, antiseptic.

ASHWAGANDHA
(*Withania somnifera* (L.) Dunal)

GYANENDRA TIWARI
Deptt. of Plant Physiology, JNKVV, Jabalpur (M.P.)

Botanical Name : Withania somnifera

Family : Solanaceae

Vernacular Names

San.-Asavgandha; Eng.-Winter Cherry, Hin.-Asgandh; Beng.-Ashavgandha; Guj.-Ason, Ashvagandha, Ghoda asor, Mar.-Asgandha; Tel.-Aswagandha, Pannerugadda; Tam.-Achuvagandi; Kain.-Ashvagandhi, Sogade-beru; Mal.-Amukkiram, Pivate; Trade-Ashwagandha.

Introduction

It is mentioned as an important drug in ancient Ayurvedic literature. Ashwagandha is an important cultivated medicinal crop of India. Ashwagandha belong to the family solanaceae. Seven types of alkaloids are found in this plant, out of which withanine and somniferine are important. Withaferin-A a Steroidal Lactone, possesses growth inhibitory and radio-sensitizing effects on experimental mouse tumors. Ashawgandha could prove to be a good natural source of potent and relatively safe radio-sensitizer/chemotherapeutic agents. The properties of ashwagnadha roots and the restorative properties of ginseng roots have led to it being called Indian Ginseng.

Botanical description

An erect, herbaceous, evergreen, tomentose shrub, 13-150 cm high. All its parts are clothed with whitish, stellate hairs. Leaves ovate, entire, thin, base cuneate and densely hairy beneath. Flowers are greenish or lurid yellow, axillary, in clusters of about 25 forming umbellate cymes, sessile or subsessile. The fruit is a berry, 7 mm across, red, globose, smooth, enclosed in an inflated, membranous, somewhat 5-angled, pubescent, persistent calyx. The fruits turn orange-red in colour when they mature. The seeds are yellow in colour and reniform in shape. The flowering season is from July to September and the ripe fruits are available in December.

Geographical distribution

It is native to Mediterranean region in North Africa. It is found wild in grazing grounds in Mandsaur and the forestlands in the Bastar district of Madhya Pradesh, all over the foothills of the Punjab and Himachal Pradesh and Western Uttar Pradesh, in the Himalayas. Other than India, it is also found in the wild in the Mediterranean region in North Africa, Spain, Island, Morocco, Jordan, Baluchistan (Pakistan) and Sri Lanka. The crop is cultivated in an area of about 4000 ha in India, mainly in the drier parts of Manasa, Neemuch and Jawad Tehsils of the Mandsaur District of Madhya Pradesh, in Punjab, Sind, Rajasthan and South India. In Karnataka, its cultivation has been reported in the Mysore district.

Medicinal uses

The drug is mainly used in Ayurvedic and Unani preparations. Withaferine-A has been receiving a good deal of attention because of its antibiotic and anti-tumor properties. The paste prepared out of its leaves is used for curing inflammation of tubercular glands and that of its roots for curing skin diseases, bronchitis and ulcers. In some areas, the warm leaves are also used for providing comfort for eye diseases. However, the roots are mostly used for curing general and sexual debility.

Its fruits and seeds are diuretic in nature. For the treatment of piles, a decoction of the leaves is used both internally and externally. The leaves are also used as a hypnotic in alcoholism. An ointment prepared by boiling the leaves, is useful for bedsores and wounds.

An infusion of the bark is given for asthma. For chest complaints, colds and chills, its decoction is recommended. The root is used to treat snakebite. It is also used in chest complaints. A paste made of the green berries with leaves and small twigs is useful for treating saddle sores and girth-galls in horses.

Soil and climate

It grows successfully in sandy loam or light-red soils (PH 7.5-8.0) with good organic matter and drainage. It prefers a subtropical dry climate. Annual rainfall of 660-750 mm is best suited its growth.

Improved varieties

Jawahar Asgandh-20 (WS-20) and WS-22, WS 134 and Rakshita.

Agro-Technique

Land preparation

Before cultivation, the land should be ploughed twice and the field should be cleaned thoroughly of weeds. If required, small canals may be prepared for drainage. About 25 t/ha of cow dung manure is also added.

Mode of propagation

It can be propagated by seeds.

Direct sowing

In this case, the seeds are sown directly in the main field by broadcasting. Since it is largely grown as a rainfed crop, the sowing is determined by the monsoon. After receiving one or two showers, the field is thoroughly prepared, divided into plots of convenient sizes and the seeds are sown during the second week of July. A seed rate of 10-12 Kg/ha is required for this method of planting.

Nursery raising

When the seedlings are to be raised for transplanting, they should be sown in well-prepared, raised nursery beds. About 5-6 Kg of seeds are required to provide enough seedlings for sowing one hectare. To avoid nursery diseases, the seeds are treated with common fungicide such as Dithane M-45 of seeds before sowing to protect the plant from seeds born fungal disease or seedling rot or wilt. The germination commences within 6-7 days of sowing and in about 10 days after sowing it is complete.

Transplanting

When the seedlings are 6 weeks old and sufficiently tall they are transplanted in 60 cm-spaced rows, 60 cm apart in well-prepared land. The optimum time of Transplanting by August-September.

Manures and fertilizers

In Madhya Pradesh, where it is grown on a commercial scale, no fertilizers are applied and the crop is cultivated on only residual fertility. Organic manures @ 10-15 tonnes/ha should be applied along with the chemical fertilizers. Ammonium sulphate @ 30 Kg/ha, when added to the soil shows positive results. The phosphorous fertilizer should be added to the field before sowing @ 18-20 Kg/ha. Nearly 10 Kg of nitrogen should be added as a seed dose to the crop 60-80 days after sowing.

Irrigation

The crop does not require any Irrigation. However, in areas where rainfall is restricted to a few months in a particular period, about 2-3 irrigations will help the plants to give optimum yield.

Interculture

In broadcast plots only hand weeding is possible where as in line intercultural implements such a way that they do not over the crops. Weeds cut or plucked also can be used for mulching. The directly sown crop is thinned at 25-30 days. Hand weeding at 30-day intervals helps to control the weeds effectively.

Pests and diseases

Seed rotting, seedling blight and leaf blight are common diseases affecting Ashwagandha. Their incidence can be minimized by treating the seeds before sowing with Captan at the rate of 3g/Kg of seeds, followed by spraying the crop with Dithane M-45 at the rate of 3 g/l of water when the crop is 30 days old. The spray should be repeated at an interval of 7-10 days if the disease is not controlled.

Harvesting

Harvesting starts from January and continues till March (150-170 days after sowing). The maturity of the crop is judged by the drying out of the leaves and the berries turning red. The entire plants are uprooted and the roots are separated from the aerial parts by curing the stem 1-2 cm above the crown.

Production and yield

An average yield of 3 to 5 qt/ha of dried roots and 50 to 75 Kg/ha seeds can be obtained.

BRAHMI [*Bacopa monniera* (L.)]

V.K. SONAKIA
Professor
Department of Plant Physiology, J.N.K.V.V., Jabalpur

Botanical Name : *Bacopa Monniera*

Family : Scrphulariaceae

Vernacular Names

Sanskrit: Nira-brahmi; Hindi: Brahmi; Bengali: Brahmi-sak; Marathi, Tamil and Malyalam: Neer-brahmi; Kannada: Nirubrahmi; English: Thyme-leaved Gratiola.

Habit and habitat

A creeping herb, commonly growing on riverbanks and marshy places throughout plains of India.

Description

A prostrate or creeping, juicy, succulent, glabrous annual herb, rooting at the nodes with numerous ascending branches; leaves simple, opposite, decussate, sessile, obovate-oblong or spathulate, entire, fleshy, obscurely veined, punctate on long slender pedicels; fruits ovoid, acute, 2-celled, 2-valved capsules.

Flowering and fruiting: Almost throughout the year

Soil and climate

The plant is found growing as a common weed in wet or marshy lowlands in plains, throughout India. The plant tolerates heavy metal toxicity.

Mode of propagation

It is very easily propagated by stem cuttings.

Cultural practices

The plant can be grown in low-lying areas of the farm, drainage trench and water storage ponds. The plants become dormant in winters and start regenerating with advent of summers. The crop can be harvested thrice, i.e. in April, June (end) and September with assured supply of irrigation. The water requirement of the crop is very high.

Medicinal Uses

- Decoction of the leaves is used in cough.
- Fresh plant is ground and applied externally on eczema and ringworm disesses.
- Aqueous leaf juice is given, twice a day, for one month to diabetic patients.
- Plant juice with black pepper is given, twice a day for three days, in malarial fever.
- Juice of the leaves (10 ml) with 10 ml of honey is given, once a day, empty stomach for 30 days in epilepsy.

Important formulations

Sleepinol, Memoryplus, Siledin, Brahmighrit, Saraswatarista, Rumanyl tablet.

Active Constituents

Brahmine, Herpestine, Monnierin, Bacosides A and B.

MANDUKPARNI [*Centella asiatica* (L.)]

V.K. SONAKIA
Professor
Department of Plant Physiology, J.N.K.V.V., Jabalpur

Botanical Name : *Centella asiatica*

Family : Hydrocotylaceae

Vernacular names

Sanskrit: Mandukaparni; Hindi: Brahma-manduki, Khulakhudi, Mandookaparni; Bengali: Thol-kuri; Gujarati: Moti Brahmi; Marathi: Karinga, Karivana; Kannada: Brahmisoppu; Telgu: Brahmi, Saraswataku; Tamil: Vallrai; Oriya: Thalkudi; Malayalam: Kodangal, Muyalachevi; English: Indian Pennywort.

Habit and habitat

A prostrate, fantily aromatic, stoloniferous perennial herb, commonly found as a weed in crop fields and other moist places of India.

Description

A prostrate, slender, herbaceous, creeping perenniel with rooted nodes and long internodes; leaves fleshy, simple with elongated petioles and sheating leaf bases, broadly cortdate, orbicular-reniform, crenate-denate, smooth on upper surface and sparsely hairy on the lower; flowers red, pink or white, in fascicled umbels; fruits oblong, dull brown, laterally compressed, pericarp hard and thickened.

Flowering and fruiting: April-October

Cultivation

Soil and climate

The plant is found in wet places, banks of streams and ponds throughout the tropical & sub-tropical regions of India.

Mode of propagation

The plant is propagated by seeds and stem cuttings.

Cultural practices

The plants grow well under shade and can tolerate heavy shade of mango orchards. One-node stem cuttings can be planted at 45 x 45 cm and irrigated. The roots develop from nodes. The planting can be done in February-March with assured irrigations and in July as rainfed crop. The leaves are harvested and dried in shade.

Medicinal uses

- Leaves are taken to improve memory and serve as a nerve tonic. 3-4 fresh leaves with 11-12 black peppers are given to children to improve the memory.
- Extract of plant is taken in fever.
- Two to three tsp of leaf paste is administered, empty stomach, for 2-3 days in dysentery.
- The plant made into a paste is applied on boils and tumours.
- The syrup resulting from the boiling of leaves, with small ginger and black pepper, is taken as cough syrup.
- Juice of leaves mixed with palm jaggery is given twice daily for 15 days as a tonic after delivery.
- Leaf juice is given for the treatment of diabetes.
- Leaf juice is rubbed on forehead to cure severe headache.

Cosmetic uses

Plant extract used as cream to improve skin texture and also in hair-oil.

Important formulations

Brahmi ark, Dabur Lodised Sarsa, diabel, Nardyl tablets, Geriforte tablets.

Active constituents

Asiatic acid Asiaticoside, Madecassic acid, Brahmic acid, Thankuniside, Centellose.

Price

Centella asiatica Rs. 16 per kg.

BACH OR SWEET FLAG (*Acorus calamus* L.)

S.K. DWIVEDI
Associate Professor
Deptt, of Plant Physiology, JNKVV, Jabalpur (M.P.)

Botanical Name : *Acorus Calamus*

Family : Araceae

Vernacular names

Sanskrit - Vacha Ugragandha, Hindi-Bach, Bengali - Bachh, Gujrati - Godavaj, Panjabi - Warch, Tamil - Vasamboo, Malyalam - Vayambu, Vayambu, Telgu - Vasa; Kannad - Baji, Marathi - Vekhad, English - Sweet flag, Trade-calamus.

Distribution

It grows wild on the edges of swamps, on the banks of ponds and river in Europe, North America and India. It is commercially cultivated in the USSR, Central Europe, Rumania, India and Japan. In India the leading states are Karnataka, Kashmir, Manipur, Arunachal Pradesh, Nagaland, Meghalaya and in the foothills of Himalaya up to 1800m. In Madhya Pradesh Vindya plateau, Narmada Son Balley, Satpura, Mekalregion and the districts of Jabalpur, Sagar, Mandla, Dendori, Shahdol are the major growing areas of this crop.

Introduction

Sweet flag (*Acorus calamus* L) is an aromatic herb belonging to the family Aeraceae. The dried pleasant smelling rhizomes is used in medicinal and as aflavourant. The rhizome contains a volatile oil with an acrid taste and a bitter substance. The volatile oil is yellowish brown and can be isolated by steam distillation.

Botanical description

It is a semi-aquatic perennial monocot herb with a thick-branched aromatic rhizomes. These rhizomes are 1.5 - 2.5 cm thick and some times up to 2 m long and horizontally immersed in the mud with short joints and large leaf scars. The leaves are narrow sword like leaf blades up to 1 meter long and are produced at the tip of the rhizome. The leaves are bright green with a medium thickening and wavy margins. The flowers are light brown in a densely packed 3 to 5 cm long sessile cylindrical spadix at the leaf bases. The fruit is a berry yellow green in color, angular and 1-3 seeded.

The economic part of the plant is the underground rhizome. The rhizomes are horizontal and jointed, with a spongy texture.

Climate

Bach is a hardy plant and is thus found growing from tropical to subtropical climates. It prefers an area with good and well-distributed rainfall throughout the year.

Soil and soil preparation

It is cultivated in almost the same way as paddy and may be grown in any part of the country where suitable irrigation facilities are available. It can be grown profitably on well-drained sandy loam to loamy soil with assured irrigation. It also prefers soil possessing high content of humus for better initial growth.

The soil is prepared exactly same as for paddy. It irrigated sufficiently and after ploughing twice; is watered heavily and ploughed again. The field is then left for few days and puddled. After that, it is levelled with a levelling board.

Propagation

The planting material of sweet flag is consists of the live ends or tops of the previous crop. At harvest the mature portion of the rhizomes is cut off, which is the marketable part and the tender portion formed by the growing end is used for replanting and propagation. This propagating material can be planted at once or stored for about a week by covering it with straw or dried leaves. If it is required to be stored longer, this can be done by keeping it in openpits.

Planting and planting time

The planting is done in the furrows at a depth of 5 cm and maintains a spacing of 30 cm between the rows. The newly sprouted rhizome propagules are planted about 30 cm apart. This is a kharif crop and can be planted in June and July. The best time for planting is second fortnight of June.

Manures and fertilizers

A basal dressing of farmyard manure @10 tonnes/ hectare should be applied to the land before planting the crop. The crop also responds significantly to the application of nitrogen and phosphorus. The recommended fertilizer doses of 125 KgN, 50 KgP_2O_5 and 60 Kg K_2O/ha. It is advisable to use one third of nitrogen at the time of planting. The second dose of nitrogen should be given after one month of planting, as broadcast and third dose should be applied after two months of planting.

Irrigation

During the winter season the crop should be irrigated at weekly interval and while in summer the irrigation should be given twice a week. The rhizomes do not go deeper than 5 cm in the soil and accordingly depth of irrigation is regulated at different stages of the crop.

Inter culture operations

Careful and regular weeding is more appropriates to obtain high yield and at weeding the growing parts are lightly pressed down into the soil.

Insect, diseases and their control

Generally this crop is free from any attack of insects and diseases. The attack of mealy bugs is reported which infest the shoot and roots and may be controlled by spraying the shoot and drenching the roots of grown plants with 0.1% methyl parathion or 0.15% oxydemetonmethyl or 0.2% quinolphos.

Harvesting and storage

After about 6-8 months the Bach crop is ready for harvesting. The field is allowed to dry partially; so that sufficient moisture is retained to facilitate the necessary deep digging. Now, the leaves start turning yellow and dry indicative of maturity. The rhizomes after harvest are cut into small pieces of 5-7.5 cm and all the fibrous roots are removed. The pieces then washed thoroughly and dried in the sun. The dried rhizomes may be stored in the gunny bags.

Yield

Under favorable condition a yield of 42 quintals / hector of dried rhizomes is obtained. The fresh rhizomes yield about 2.5-3.5% essential oil. The fresh aerial parts yield 0.10% oil.

Chemical constituents

The rhizomes; roots and leaves yield an essential oil; the important constituents of the Indian oil are asarone and its b-isomer; other constituents are a and b pinene, myrcene, camphene, P-cymene, camphor and linalool, sesquiterpenic ketornes like asarone, colamone, calcone and acolamone etc. Root yield acoricacid and choline.

Medicinal uses

The dried pleasant smelling rhizomes and its essential oil both are used as aromatic bitter in carminative composition and for bronchial troubles. It is a good sedative so that the extract is used for epilepsy; insanity and as a tranquilizer along with mushkbala (*Valeriana jatamansi*) and jatamansi (*Nardostachys grandliflora*).

It is an ingredient of an Ayurvedic preparation. BrahmiBati (Budhivardhak) which is indicated in epilepsy, coma, hysteria and in cases of mental retardation; the same uses are prescribed for Acorus, containing Unani drug Majun Baladur.

Bhringraj [*Eclipta alba* (L) Hassk.]

S.K. DWIVEDI
Associate Professor
Deptt. of Plant Physiology, JNKVV, Jabalpur (M.P.)

Botanical Name : *Eclipta alba*

Family : Compositae

Vernacular names

Bhringarauah, Telarajah (san.), Bhamgra, Mocakand, Babri, Bhringraj, Ghamra (Hin.), Kesutthe, kesraj (Ben.), Kannunni, Kayyonni, Kayyuni (Mal.), Kayuntakara, Kaikeri (Tam.), Garagaddasopu, kadiggagaraga (Kan.), Guntugulijeran, Galagara (Tel.), Bhangaro (Guj.), Maka (Mar.), Trailing eclipta (Eng.), kadi-el-bint (Arab).

Botanical description

Bhringraj is a prostate annual herb with erect main stem, 10-15 cm tall and trailing, numerous branches, rooting at nodes. Leaves are opposite with hairs on both surfaces. Flowers are white in heads. Floral heads are 6-8 mm in diameter and solitary. Fruit is an achene, compressed and narrowly winged. Sometimes *Wedelia calendulacea*, which resembles *Eclipta alba* is used for the same purpose.

Distribution

The eclipta is widely distributed in the warm humid tropics with plenty of rainfall. It grows commonly in moist places as a weed all over the Plains of India.

Climate

It may be grown in almost all types of climates. It may be cultivated in both kharif and Rabi seasons.

Soil and soil preparation

It is a hard plant and can be cultivated on a wide range of soils under various agroclimatic conditions. The field is thoroughly prepared and levelled.

Sowing time and method

The seeds are dibbled in rows preferably 20-30 cm, apart, keeping plant to plant spacing around 5-10 cm. The crop is sown in June or July in Kharif and October to December in Rabi season for good harvest.

Harvesting and storage

The crop duration is about 4-5 months in both kharif as well as in rabi seasons. The first harvest (cutting) may be done 60-75 days after sowing and thereafter with an interval of 1 month. Leaves are sun dried during kharif and in shadow during Rabi season. The dried leaves should be stored in gunny bags.

Production and yield

The crop is harvested three times during crop duration which yield about 10 quintals dried leaf and 2-3 q seed per hectare.

Chemical constituents

The dried leaves of *Eclipta alba* have been reported to cortaiu wadelolactore, a complex coumarin and its derivatives dimethyl wadelolactorie - 7 glucoside and norwadelolactorie.

The root contains polyacetylene-substituted thiophenes and leaves have been reported to contain terthienylmethanol. The aerial part of plant has been reported to contain a phytosterol, b-amyrin in the n-hexane extract and luteolin-7-glucoside, b-glucoside of phytosterol, a glucoside of a triterpenic acid and wadelolactone in polar solvent extract.

Hentriacontanol and heptacosanol are also present in roots. The plants also contain nicotine and nicotinic acid.

Medicinal uses

The Bhringraj *(Eclipta alba)* is mainly used in hair oils, but it has been considered a good drug in hepatotoxicity. In hair oils, it may be use along with Mandookparni (*Centella asiatica*) and anola (*Phyllanthus emblica*). It is used to prevent abortion and miscarriage and also in cases of uterine pains after the delivery. The juice of the plant with honey is given to infants for catarrh and with castor oil for expulsion of worms. For the relief of piles; fumigation with Bhringraj is considered beneficial. A decoction of leaves is used in uterine haemorrhage. The paste prepared by mincing fresh plants has got an anti-inflammatory effect and may be applied on insect bites, stings, swellings and other skin diseases. In Ayurveda, it is mainly used in hair oil while in Unani system; the juice of plant is used in 'Hab miskeen, nawaz' alongwith aconite (*Aconitum spp.*), Jayphal (*Croton tiglium*), Triphala, black piper (*Piper nigrum*), Pippali (*Piper longeum*), Zinger (*Zingiber officinale*) and minerals like mercury, sulphur, arsenic, borax etc. for various types of pains in the body. It is also a constituent of 'Roghan Amla Khas' for applying on the hair and of 'Ma Jun Murrawah-ularwah'. The doses of plant juice are 5 to 10 ml and powder is about 1-3 g.

KALMEGH (*Andographis paniculata*)

V.K. SONAKIA
Professor
Department of Plant Physiology, J.N.K.V.V. , Jabalpur

Botanical Name : *Andrographis paniculata*

Family : Acanthaceae

Vernacular names

Sanskrit: Bhunimba, Kirata;
Bengali: Kalmegh;
Gujrati: Kariyatu; Marathi: Olikiryata;
Oriya: Bhuinimba; Kannada: Nelaberu;
Tamil: Nilavembu; Telgu: Neelaveemu;
Malayalam: Kiriyattu;
English: The Creat;

Habit and Habitat

Herb grows abundantly in moist, shady, waste grounds and sometimes in dry forests.

Description

An erect annual herb; stem dark green, quadrangular with longitudinal furrows and wings on the angles of the younger parts, slightly enlarged at the nodes; leaves glabrous, lanceolate; flowers small, in lax spreading axillary and terminal racemes or panicles; capsule linear-oblong, acute at both ends, seeds numerous, sub-quadrate, yellowish brown.

Flowering and fruiting: October-December

Cultivation

Soil and climate

It grows in moderately fertile and well-drained loamy or sandy soil. The plant grows in a wide range of temperature in dry to moist places, mainly in plains.

Mode of propagation

The field should be well ploughed and seeds sown at the rate of 500-600 g in a nursery bed for producing seedlings for one acre area in May-June. When the seedlings are 6-8 cm, transplanting is done by keeping the space at 45 x 30 cm between plants and the rows. Light irrigation is required after transplantation.

Cultural practices

The crop needs frequent irrigations after transplantation. With the onset of monsoon, the rain meets the requirement till September. Thereafter, irrigation may be required once or twice a week till their maturity in December. Total 3-4 weedings are required for the crop season. At least two hoeings before the first harvest are important.

The whole plant constitutes the drug. Harvesting should be done at 10 cm above the ground level. Two or three harvests of leaves can be taken during the crop season. To retain the active principles, the leaves should be dried at a suitable temperature by spreading the leaves in thin layers on the ground for 36-48 hrs.

Status in forest flora and constraints in production

Andrographis exists in forest flora of Uttar Pradesh, Assam, Madhya Pradesh, Jammu and Kashmir, Tamilnadu, Kerala and Orissa state there is exploitation of this plant from the forest under nontimber minor forest produce and sold through cooperative societies.

Constraints

Andographis is recently domesticated. Thus crop maturity in nonsynchronous that lead to lot of problems the time of crop harvesting there is need to have selection and / or management for synchrony for crop maturity.

It is yet to be finalized by physiologist which plant part has maximum alkaloids for high active ingredient containt and to select morphological indice for harvesting that part.

Medicinal uses

- The juice of the bruised leaves is mixed with honey in proportion and is given to children in fever and cough. Nothing sour is allowed during the period of treatment.
- The whole plant is dried in the shade, made into powder and given with water for malarial fever and other intermittent fevers.
- The decoction of leaves, with common salt and black pepper is taken in fever.
- The extract of plant, mixed with common salt is taken in cholera and dysentery.
- The mixture of root paste and black salt in water is taken as stomachic.
- Whole plant of 'kalmegh' and stem bark of 'neem' (*Azadirachta indica* A. Juss.) and 'kureya' (*Holarrhena antidysenterica* (L.) Wall. ex G. Don.) are taken in equal quantities, pounded well and mixed in water about 6 times by volume. A red-hot brick is then dropped into it and filtered. The extract thus obtained is given internally for jaundice. Dose: 5 ml, 3-4 times a day for 3 days.
- Dried root of 'kalmegh' with 'haldi' (*Curcuma longa* L.) in equal quantities is made into a paste with water. This paste is applied for itching and skin rashes.
- Root of the plant is made into powder, 3 g powdered root given with water to children once a day as an anthelmintic.
- The infusion of the whole dry plant is taken daily empty stomach to treat sluggish liver, neuralgia, dyspepsia and general debility.

Important formulations

Hepargen, Siomethionine, Livopep, Kalmegh navayas, Livotrit liquid, SG-I Switradilepa, Tefroli

Active constituents

Andrographolide, Neoandrographolide, Deoxyandrographolide and Andrographiside.

SAFED MUSLI
(*Chlorophytum borivillianum*)

GYANENDRA TIWARI
Deptt. of Plant Physiology, JNKVV, Jabalpur (M.P.)

Botanical Name : *Chlorophytum borivillianum*

Family : Liliaceae

Vermacular Names

San, -Musali, Hin.-Safed musli, Kulai, Mar.-Kuli, Safed musli, Guj.-Gholi-musali, Janjaria, Tam.-Tan-ureta, Tel.-Kushelli.-Mal,-Shedeveli.

Introduction

Tubers are white, and hence it is called as safed musli. Musli tubers available in the markets are normally collected from tribals from hilly areas. Safed musli is an important Ayurvedic medicinal plant. The small seedlings of musli are found in forests during rainy season.

Botanical description

Leaves look like that of garlic but slightly yellowish, and white flowers with 6 petals are arranged on the flowering stalk, which emerge from the center of the plant. About 20-25 flowers on the flowering stalk appear in July. The seed is very small, black and enclosed in flowering boles. In one bole there are about 10-12 seeds.

Geographical distribution

The plant is observed in the valley of Himalaya, Satpuda, Vindhya, Arawah and in hilly areas of Bihar of Assam. It is also found in the parts of Rajasthan, Gujarat in some parts of Maharashtra.

Medicinal uses

It is useful on *itla* and *uata*. It is explained as *urishya, dhatuvardhak, sheet, madhur, rasayana, slightly* bitte, healthier and energetic in Ayurveda. It is effective on *fatigueness, daaha* and in blood purification. It is useful in certain diseases like renal calculus, dhupani, sangrahani, leucorrhoea. It is lactating, energetic to heart.

Chemical constituents

Tubers are of great medicinal value containing the steroid sapoginine (1-2%), proteins (10-12%) and calcium.

Soil and climate

The loamy, gravelly soil with rich organic matter and well-drained soil with pH ranging between 6.5 to 7.0 is suitable for growth. Good amount of soil moisture during growing season favours freshly root development. Dry weather is favourable for its growth. It can survive at low temperature. It can be grown under warm humid climatic conditions.

Improved varieties

RC-2, RC-16,RC-36,RC-20RC-23 and RC-37.

Agro-Technique

Land preparation

The land should be ploughed and harrowed so that there should be proper aration. Weeds like *Cyprus rotundus, Cynodon dactylon* should be killed completely. After that the broad bed furrows 90 cm broad and 15 cm height should be prepared. Water channeld should be prepared for eachf bed. Well-decomposed FYM about 15-20 tonnes/ha should be mixed in the soil properly before planting, or the vermicompost can be used.

Mode of propagation

It is propagated by seeds and tubers. But the germination percentage by seed is very less and it takes about 18 months for harvesting tubers. Hence, the sprouted tubers are economical for plating.

Sowing

The bunch of tubers should be removed from soil tubers are separated from the bunch in such a way that some portion of crown should remain attached to each tuber. The number of tubers varies from plant to plant and on an average 5-30 tubers/plant. Tubers separated should be stored in gunny bags filled with mixture of fine sand and silt with moist condition under shade in June, tubers get sprouted naturally and these are used for planting.

Planting

Planting should be done on the onset of monsoon, preferably in June. The soil should be moist. The planting should be done at the spacing of 6-8 cm between tubers while distance between rows should not be less than 30cm. Tubers should be placed deeply in the soil.

Irrigation

Irrigation should be given as and when required during growth period to avoid any water stress. Sprinklers may be used if available for irrigation purpose.

Interculture

Weeding should be done by hand. As the plants of musli appear like that of grass, there is possibility of getting removed these musli plants along with grass. The field should be kept weed free.

Harvesting

Tubers are ready for harvesting after 6 months of planting. Harvesting is done in December, January, when the matured leaves drop down. This is the best indication for harvesting tubers. The tuber bunches are removed by digging the soil.

Bunches removed from soil should be thoroughly washed with water. Such clean bunches are dried in shade for 2 days. The skin on the tubers should be removed by pressing the tubers by holding between 2 fingers. Such peeled musli should be again washed with water and dried completely in shade. The hard dried musli is marketed.

Production and yield

Fresh tubers 3.0 to 3.5 tonnes/ha or dried tubers 0.6-0.65 tonnes/ha.

TULSI (*Ocimum spp.*)

V.K. AGRAWAL
Professor and Head
Deptt. of Plant and Physiology, JNKVV, Jabalpur (M.P.)

Botanical Name	:	*Ocimum spp.*
Family	:	Labiateae
Common Name	:	Tulsi

There are about 160 species in this genus and is distributed in warmer parts of the hemispheres from sea level 1800 m. Ocimum species are cultivated commercially in France, England, America and India. These are shrubs, or herbs, perennial or biennial in habit. Plants are much branched with quadrangular stems and twigs, strongly scented.

Uses

1. Many species of ocimum contain various economically important essential oil used in perfumery and cosmetic industries.
2. The major constituent in a Ocimum oil includes, linalool, geraniol, citral, camphor, eugenol, menthy chevicol, safrol, thymal, methycinnamate, etc, that can be harnessed to yield many commercial products.
3. Ocimum species are used as herb and find diverse uses in indigenous system of medicine because many species has the antifungal, anti-bactericidal and insecticidal properties.

The genus can be broadly divided into two main groups viz basilicum group and sanctum group based on the basis of following differences:

S.No.	Particulars	Basilicum Group	Sanctum Group
1.	Habit	Species are mostly herbaceous perennials	Biennials or triennals or woody perennial under shrubs.
2.	Flower Characters	Bracts are petiolate flowers more conspicuous	Bracts are sessile Flowers are small
3.	Seed	Mucilagenous	Non-mucilagenous

Floral characters of all 0cimum species are similar in basic structure. Flowers are protandrous. The duration between opening of flowers, dehiscence of anther and maturity of style varies between species to species. Flowers are mainly entomophilus, thus they are mostly outcrossing types.

In India *O. amricanum, O. basilicum, O.canum, O. gratissimum* and *O. sanctum* are distributed, *O. gratissimum,* is mostly cultivated in north India and *O. sanctum* being considered as sacred plant by Hindus and is cultivated in almost every part of India.

Climate and soil

As it is a tropical to subtropical plant which prefers fairly high rainfall areas with humid conditions. Long day and high temperature condition enhance growth and higher oil production. Partial shade is not beneficial as it affect the oil content. It can tolerate drought and frost. Ocimum can be grown in rich loam to poor/laterite soils. It can also come up in saline and alkaline soils and also in moderately acidic soils. Proper drainage is however essential as water logging result in poor growth.

Field preparation

The land is well prepared with two or three ploughing until a fine tilth soil is obtained. FYM may be applied before 2nd or 3rd ploughing. The crop is raised from seeds and can be grown annually from the middle to the end of September by direct sowing or transplanting.

Varieties developed by RPL, Jammu.

S. No	Name of The species	Variety	Fresh herb yield (t/ha)	Oil yield Kg/ha	Oil recovery (F.W. basic %)	Major constituents	Other chemical compound
1.	*O. americanum*	RRL-02	40-42	175 TO 200	0.44	Methyl chavicol	Linalool Geraniol Borneol Edesmol
		RRL-07	40	400 Kg	0.44	Methyl Chavicol (70-75%)	Caryophyllene-oxide, Linalyl acetate Y-cadenine 1,8 cineol terpenol Lemonene Ocimene
2.	*O.basilimum*	RRL-07	40	200	0.50	Citral (75-80%)	Pinene Cineole Geraniol Linalool
		RRL-011	50	220	0.50	Linallol (50%) and Methyl chavicol (35%)	Camphor, Methyl chavicol, Eugenol Sesquiterpene Methycinnamate Ocimene leaves contain Protein Carbohydrate, Volatile oil Fixed oil, Cellulose, Mineral matter vit. A,B and C
3.	*O. canum*	RRL-01	40	200	0.5	Linalool	Linool, Campher pinene, Elemal, Methyl chavicol Geranylacetate, Linalyl acetate, Tricycline, Myrcene Ocimene and Boranyl actate
4.	*O. gratissimum*	RRL-08	40	160	0.40	Eugenol (75-80%)	Ocimene, Myrcene, Methyl chavicol
5.	(*O.gratissimum* O. x o. viride)	RRL-15	40	250	0.50	Eugenol (75-80%)	Carveol, Caryophyllene oxide, α pinene, Limonene, Phellandrene, Citral
6.	*O. viride*	RRL-08	42	200	0.50	Thymol (65-70%)	Thymol, d-imonene,
		RRL-09	42	200	0.50	Thymol (65-70%)	α-γ-terpenol Phellandrene myrcene, 4-terpenol α-terpinol Geranylacetate caryophyllene

(i) Direct sowing: Seeds (200-250 gm per ha) are mixed with sand to ensure an even distribution. Before sowing the field is marked into rows 50-60 cm apart. Seed is sown by hand or drilled & covered with soil by taking care that seed remains at a depth of about 2 cm. in the soil. Seeds sown deeper fail to germinate. The field is irrigated after 24 hours depending on soil moisture. Germination starts within 10 to 15 days. After 20-25 days when the seeding are 10-15 cm tall first weeding and thinning if necessary is done.

(ii) Transplanting: Seeds sown in the nursery beds towards end of March in raised bed. Germination is profuse and complete in about 10 days. Seedlings with 4-6-leaf become ready in 6-7 weeks for transplanting. Seedling may be transplanted at a spacing of 60 x 60 cm. The fields are irrigated both prior to and after transplanting.

Fertilizer application

An application on 20 kg N and 15 g P_2O_5 and K_2O each as basal dose and 20 kg N per hectare after one month from planting give good herbage and oil yield.

Harvesting and distillation

From sowing to first harvesting, it takes about 90-100 days in case of direct sowning and 75-90 days in the transplanted crop. The plants are cut 20-25 cm above ground level in the first year, 20-30 cm level in second year and 35-40 cm in third year. Close cutting 20 cm will be injurious to plant immediately after the harvest the field has to be irrigated. The second and third harvesting are taken at 50-60 days interval. Harvesting is generally done on a bright sunny day for better quality of oil. Oil is obtained by steam distillation method. Anhydrous sulphate 3% is added in the oil for storage. The oil is stored in colored bottle or aluminum pot.

Yield

On an average 25 to 30 tonnes of herbage yield and 150 kg of oil yield can be obtained per hectare in the first year and in subsequent years it may even go up. The oil yield depend upon the species grown and many other factors such as stage of harvest, season etc.

Net profit

A net Profit of Rs. 20,000-25,000 can be obtained per hectare/year from ocimum cultivation and oil extraction.

CHANDRASUR [*Lepidium sativum* (L.)]

R.S. PATEL
Associate Professor
Deptt. of Plant Physiology, JNKVV, Jabalpur (M.P.)

Botanical Name	:	*Lepidium sativum*
Family	:	Cruciferae
Common Name	:	Chandrasur, Halo

Description of plant

Height of Chandrasur plant is 30-75 cm. Whitish and seeds redish like as mustard seeds. Other names of Chandrasur are Asadiya, Aariya, Halim, Hali, Aaliya and Aasaliya etc.

Climate

Chandrasur crop grown in warm temperate, cool dry weather, (Winter season). Chandrasur mainly grown in Uttar Pradesh, Rajasthan, Gujarat, Maharastra and Madhya Pradesh.

Soil

Chandrasur crop require well drained sandy loomy soils, consist 6.5 to 7 pH.

(1) Field preparation -

Field must be level and pulverized by ploughing and harrowing.

Variety

JL - Chandrasur.

Seed rate

8-10 Kg/ha.

Sowing time

For higher production Chandrasur crops sown in between 15 Oct. to 15th November.

Sowing method

Chandrasur sown by broad cost or in line (Line to line distance 30 cm) seeds must be sown shallow otherwise the germination are affected.

Manure and fertilizers

(a) Mix the FYM 10-20 tones/ha at the time of field preparation.

(b) Chemical fertilizers: -

Nitrogen	-	50 Kg	Per hectare
P_2O_5	-	50 Kg	Per hectare
K_2O	-	30 Kg	Per hectare

Note – Fifty percent dose of Nitrogen (25 Kg/ha) given at the time of sowing and 25 kg/ha given after 30 days of sowing before irrigation.

Irrigation

Irrigate the field immediately after sowing for germination. Second irrigation after 20-25 days of sowing. Third irrigation at the time of milking stage.

Inter culture

Weeding require 20 days after sowing if weeds are in field.

Diseases

1. Powder mildew diseases are observed then Sulfur dust are used.
2. Controlled the aphides by spraying of Indo Sulfan insecticide 1 ml/litre water.

Harvesting

Chandrasur crop mature in 90-100 days, when leaves are yellowish and seeds turn in red colour it means crops are matured. Chandrasur crops are harvested and thrested out.

Yield

Seed yield 10-15 Qt./ha.

Uses

Salad, Green vegetable, Gum from seed, for mothers and dairy animals for increasing the milk. Animal feeds, for digestion, Eye tonic, Urine trade, Gynaecological diseases, Babaseer.

Chemical composition

Keroteen and Glycosinolate, Bejil or thaosaynet, Synapik are present in plants. Acid and Sinapin are found in seeds.

MENTHA (*Mentha arvensis*)

V.K. AGRAWAL
Professor and Head
Deptt. of Plant and Physiology, JNKVV, Jabalpur (M.P.)

Botanical Name : *Mentha arvensis*
Family : Labiatae
Common Name : Pudina

The mint species, yielding aromatic oils, comprises of about 25 species. Out of which five are commercially cultivated in India.

1. Japanese mint	(*Mentha arvensis L.*)
2. Pepper mint	(*M. piperata*)
3. Common or Spearmint	(*M. spicata L.*)
4. Scotch spearmint	(*M. cardiaca Baxer*)
5. Bergamot mint	(*M. citrata Ehrh*)

Mint cultivation is done in India since more than three decades. The area occupied by mints in India is about one lakh hactare with production of 11,000 to 12,000 tonne oil valued to Rs 125 crores. Uttap Pradesh, Punjab and Haryana are the main area where mint cultivation is done on large scale. Owing to diversified uses, the demand of their oil is ever increasing. It is being used in toothpaste, pan masala, hair oil, chewing gum, pain balm, face cream, shaving lotion, perfume and cough syrup.

These are erect or decumbent aromatic perennial herbs and the stem are mostly square in cross section. Leaves are opposite, sessile or petiolated. Flowers are densely born in terminal spike or head, many flowered.

Japanese mint (*M. arvensis*)

Distribution

This species grow wild all over wet lands in China and Japan and is commercially cultivated in Japan. China, Taiwan, Thailand, India and Brazil.

Botany

It is much branched vigorous plant grow up to 1m tall. Entire plant is covered with soft tomentum and bear glandular or non-glandular hairs. Leaves are broad and lanceolate to oblong in shape. The plant has a creeping habit and spreads in all direction. The stem is rigid, pubescent and highly branched. The leaf surfaces on both the sides appear hairy and have glandular trichomes. The essential oil is synthesized in traichomes. The verticillate inflorescence is called cyme. The flowers are small and purplish in colour.

Uses

Menthol is used as a flavouring agent, anti-septic, coolant and as a carmative. The spent grass is useful in making pulp for manufacture of hard boards and papers.

It is used cough drops, mouthwashes, flavouring tobacco, chewing gum and pan, scenting cigarettes and also for cosmetics.

Soil and climate

Mint plant can be grown in wide range of soil. Well drained deep soil with moisture retentive capacity and rich in organic matter is ideal for its cultivation. Plant prefers humid growing conditions and a soil pH range of 6 to 7.5. Long day length during the period of growth of these plants contributes towards high oil yield.

A temperature range of 20°C to 30° C, during growing as well as harvesting period and well-distributed rainfall of about 100-110 cm is ideal climate for growing Japanese mint. In India this is grown as annual crop. It can be grown in tropical and subtropical climate.

Land preparation

The field in which the mint is to be cultivated should be free of weeds. Soil should be ploughed and harrowed thoroughly to prepare a fine seed bed. Any stubles should be removed before planting.

Planting material

The plants are propagated by runners, suckers, stolons and through seedlings. Multiplication through stolon in commercially practiced. Stolons are under ground stem, light cream coloured smooth and juicy which are formed at the end of creeping roots stocks during winter. These planting material are prepared by fresh planting of rooted or tops of mentha plant at spacing of 30 x 15 cm in August. The suckers and other vegetative organs are ready for planting in the month of December to January.

The ideal time for planting of menthol mint is second fortnight of January to February end. In later stage (March-April) planting of menthol mint by transplanting of seedling raised in nursery can be done. Before planting, the suckers may be treated with 0.2% solution of any contact fungicide like capton for two minutes. Stolons are planted in shallow furrow (7-10 cm deep) opened at 45-70 cm apart. After planting furrow are covered with soil. On an average 350 to 500 kg planting material are needed for one hectare of land.

Manure and fertilizer

Farm yard manure @25 tonnes per hectare and a fertilizer dose comprising of 120 kg N, 50 kg P_2O_5 and 60 kg K_2O per hectare is recommended. Phosphorus and Potash along with one-third of Nitrogen are given as basal dose. While rest of the nitrogen is applied in to equal doses at 40 days after sprouting and another after the first harvest of herbage.

Varieties

S.No.	Variety	Herb Yield (q/ha)	Essential oil yield (q/ha)	Menthol %	Source
1.	MAS-I	200	125	82	CIMAP Lucknow
2.	MAS-2	250	150	81	CIMAP Lucknow
3.	Shivalik	300	180	75	CIMAP Lucknow
4.	Gomti	400	140	75	CIMAP Lucknow
5.	Himalaya	400	140	75	CIMAP Lucknow
6.	Hybrid-77	500	350	81.5	CIMAP Lucknow
7.	PRL 18/3	450	300	85	RRL Jammu
8.	Kosi	400	300	80	CIMAP Lucknow

Irrigation and drainage

Mentha crop requires considerable moisture distributed throughout the growing season. The crop requires three irrigations after the monsoon during September, October and November. During summer it may be necessary to irrigate the field at weekly intervals.

Water logging should be avoided and adequate drainage should be provided during the standing crop.

Interculture and weed control

Weeding is one of the important intercultural operation since yield of oil depends largely on the extent of its freeness from weeds. First hand weeding is required after 5-6 weeks of planting. This is repeated once or twice depending upon nature and extent of weeds. Light hoeing with the help of spade or hand hoe is more useful in increasing the production of mint.

Application of mentha spent or citronella waste is also helpful to suppress the weeds up to some extent. The weeds can also be controlled through the use of herbicides after 2-3 days of the planting.

Insect, pest and disease control

Plants are attacked by the diseases like wilt, rust, powdery mildew, leaf blight and stolon rot. These are also affected by the different kind of insects viz. cutworms, termites, caterpillars beetles and leaf rollers. Timely spray of insecticides and fungi sides is very important to keep plant away from diseases and insect attack.

Harvesting and distillation

The crop planted during December-January will be ready for harvest by second week of May. Second crop will be ready for harvesting sometimes in August. Third harvest may also be taken in October. Ideal time of harvesting is stage when the lower leaves starts yellowing. Entire above ground part is harvested during bright sunny days by giving a cut at 10 cm, above ground. Generally first crop yield more and the subsequent harvests give a comparatively lower yield. After the third harvest by November crop enters dormancy during winter month.

After harvesting, the herb is shade dried for few hours and then distilled. Sun dried herb contains less oil than shade dried herbs. The yield and quality of oil also depends on drying process and time taken between drying, distillation the efficiency of distillation unit.

Yield

Depending upon the management of crops the yield of fresh herb is about 20 to 40 tonnes per hactare, from which 150-200 kg of oil may be obtained.

Quality of oil and chemical composition

Japanese mint oil is colourless to pale yellow with characteristic mint odour. The oil content is 0.4 to 0.5 percent on fresh weight basis. The primary constituents of oil are menthol and menthone. Other chemical constituent identified are isomenthone, limonene, cincole, octanol, methyl acetate, neomenthol, pepritone and pulegone.

ISI requirement for the oil menthol mint

S.No.	Characteristics	Requirement Type-I	Dementholated Oil Type-II
1.	Colour and appearances	Colourless pale, yellowish or greenish yellow	Pale yellow Type-2
2.	Odour and taste	Stong minty herbal, followed by cooling sensation	Intense, mint herbal, followed by cooling sensation
3.	Relative density at 27°C*	0.8773 to 0.9123	0.8923 to 0.9123
4.	Optical rotation	–35°C to –45°C	-20°C to –40°C
5.	Free alcohols as I-menthyl acetate, percent by mass	3 to 5	5 to 30
6.	Esters, as menthylacetate, percent by mass	3 to 5	5 to 30
7.	Total alcohols as menthol percent by mass	64	64
8.	Ketones as menthone percent by mass	5 to 20	20 to 40
9.	Refractive index at 27°C+	1.462 to 1.4362	1.4519 to 1.45632

* The correction factor for each degree Celsius change in temperature in 0.000238

* The correction factor for each degree Celsius change in temperature in 0.00064

Net profit

A net profit of Rs. 25,000-30,000/- per hectare can be obtained from Japanese mint cultivation and distillation.

Common or spear Mint (*M. Specata L.*)

This species is native to Europe, grows to height of 60 cm with erect stem. It is cultivated commercially in America, China and India. It is a hybrid between *M. Longifolia X M. rotundifolia*. Leaves are nearly sessile, ovate-lanceolate, serrated, light green and prominently veined. Flowers are purplish in colour, minute appear in whirled leaf spike 15 cm. Long. Some high selections in India are Punjab spearmint-I, II from CIMAP viz. MISS-I and MISS-5.

Plants require warm, sheltered and shady situation. A rich, moist, clay soil is ideal for its cultivation. Plants are propagated by cutting of basal shoots. Planting in site is done at 30 cm distance and land should be prepared by repeated ploughing and harrowing. Around 20 tonnes of

FYM is added to soil as a basal dressing. They respond well with NPK fertilization. For one hectare 125 kg of nitrogen is recommended. Irrigation weeding and hoeing operation in time add to yield and quality of oil. Fresh herb yield per hectare is 20 tonnes, which may give 60 kg of oil.

Maximum quality of oil is obtained from herb during flowering stage. On steam distillation of leaves and flowering top, 0.25 to 0.30 percent of volatile oil is obtained. The components of essential oil of *Mentha spicata* are cymene, terpinolene, b-bourbonene, dihydrocarvone, caryophllene, dihydrocarvyl acetate, terpinyl acetate, carvon and hapthalene.

Spearmint oil is used in toothpaste, chewing gum and confectionery. It is rich in carvone, a digestive and gastro-stimulant compound.

Scoth Mint (*Mentha Cardiaca Baker*)

It occurs in America. Flowering top contain essential oil, which is used as a flavour and carminative. Organic rich, moist soil is good for its cultivation. Plant is propagated through cutting and division from runners. It is also good source of spearmint oil and it contains 65 percent of carvone. The thin paste of leaves is applied to rheumatic and painful swollen joints. It is antiseptic, stimulant, administered in digestive complaints and fever. It is also used to treat diarrhoea bronchitis and dysentery. It has carminative properties.

Pepper Mint (*Mentha Piperata L.*)

It is native of Mediterranean countries. It is cultivated in the temperate region of Europe, Asia, North America and Australia. In India there is a scope to cultivate this crop in the hills of temperate and warm temperate zones like Nilgiris and upper Pulney hills of Tamilnadu besides in plain of North India. The oil is used in the pharmaceutical and flavouring industries.

Uses

The oil is used in dental cream; toothpastes mouth washes, chewing gums, alcoholic liquors and confectioneries. It possesses antiseptic and carminative properties. The oil is said to have local anesthetic influence, stimulates elimentary canal and also knows as gastric stimulant.

Botany

Peppermint is a perennial, glabrous, strongly scented oil yielding plant. It is a triple cross between *M. sylvestris x M. rotundifolia x M. ruatica*. These are much branched plants, redish or black stemmed and attain a height of 60-cm. Leaves are petiolate, ovate-oblong, acute, serrately and roundly crenate at base, glabrous and short stalked. Flowers are purple and spikes are loose and long.

Varieties

There are three varieties viz. 'Black Mitcham, White mint and DISS-5 developed by CIMAP Lucknow, Nanital. One exotic introduction EC 41911 is also recommended as it has higher menthol content.

Climate and soil

Peppermint requires temperate to subtropical climate. It prefers long dry conditions and requires an annual rainfall about 100-120 cm. It can withstand frost also.

It grow well on a wide range of soils but thrives best in deep well drained soil, rich in organic matter and has high humidity. Neutral to slightly acidic soil is considered good for its cultivation.

Land preparation

The field should have a fine tilth, which can be obtained by through ploughing and harrowing.

Planting

These plants are propagated by suckers. These suckers are collected from the previous crop and planted at a spacing of 45 x 30 cm in rows. The planting can be done from

December to March. Second week of February is considered best for herbage yield.

Manure and fertilizers

While preparing soil 25 tonnes of FYM is to be incorporated in the soil. In the plains of north India, the crop requires to be irrigated at fortnight interval from March to June, however, a week prior to harvest, it has to be stopped.

Weed control

Weeding and hoeing operations are to be attended at regular intervals. Pre-emergence weedicide application of terbacil @ 1.5 kg a.i. / ha and application of propanil @ 1.75kg a.i./ha effectively control the weeds in the field.

Harvesting and distillation

To get good quality oil, the herb material should be harvested at full bloom stage. Excessive drying of leaves in the field should be avoided. Under Indian condition optimum time for first cuttings is between the middle of June to middle of July. The second and third cutting are taken 65 to 85 days after first and second cutting respectively.

Oil is distilled through steam distillation. Care must be taken to properly distill the oil, as improper distillation gives a strongly burn odour to the oil. The recovery percent generally varies from 0.25 to 0.30 percent of the oil.

Yield

On an average 10 to 12 tonnes of herbage yield per hectare can be obtained in first year, which on distillation may yield 55 to 80 kg of oil. Yield after one year reduces very much, hence not economical.

Chemical constituent

Chemical compound isolated from peppermint oil are Acetadehyde, Dimethyl sulfide, Isovaletic aldehyde, free acetic acid, Idoamyl alcohol. Amyl alcohaol, L-pinene, 1-menthone, 1-menthol, Menthenone, Menthyl acetate, Peperitone, Menthyl isovalerate, Menthyl ester, Cadinene lactone, Menthofuran, M thymol, carvacrol, Jasmone, and 1-caryophyllene.

Bergamot mint (*M. Citrata*)

The species is naturally distributed in Europe. The species is commercially cultivated in China, Taiwan and India.

Botany

It is a hybrid between *M. aquatica x M. virdis* and attain a height of 30 cm. Leaves are petiolate, ovate, broadly rounded at the base or sub-cordata, deep green in colour or yellow variegated. Flowers are reddish purple in colour and appear in round terminal heads. Leaves when pressed release fragrance of orange.

Climate and soil

It is a crop of temperate region but can be commercially grown in sub-temperate and subtropical regions.

Well-drained sandy loam soil, rich in nutrient content with high moisture retentive capacity and humid growing conditions are suitable for cultivation of these plants. The field is thoroughly ploughed and harrowed to have fine tilth.

Manure and fertilizers

Manure and fertilizers required per hectare are 20 tonnes FYM, 150 kg N, 40 kg each of P_2O_5 and K_2O.

Irrigation and weed control

Adequate irrigation, timely weeding and hoeing are essential operation for Bergamot mint cultivation.

Varieties

CIMAP, Lucknow has developed a new variety, 'Kiran' by mutation breeding which is tall and vigorous in growth.

It produces about 239 kg Oil/ha with 48% linalool and 37 % Linalyd acetate. Punjab Agricultural University, Ludihana has also developed a variety named Roy-I.

Planting

The crop can be propagated vegetatively by runners, suckers and stolons. Runners are the best propagating material and in that case, each cutting should have at least one leaf for successful rooting.

Optimum time of planting is January while in the hill it is February to March. Spacing of 40 x 40 cm or 40 x 30 cm is recommended.

Harvesting

Harvesting of crop for extraction of essential oil is done when the plants come to flower. In most of the places the first cutting is taken during May to June, second harvesting is taken during August/September and third harvest during October/November. To boost up growth, application of GA (200 ppm) is recommended.

Yield

Yield of fresh herb per hectare varies between 20-25 tonnes which may give 100 kg of oil. By steam distillation of fresh herb 0.4 to 0.5 percent of oil can be obtained.

Quality and chemical composition

The oil is pale yellow is colour with fruity citrus odour. The chemical constituents of the oil are linalool, linyl acetate, ester, free acid and aldehyde. Linalyl acetate is principal constituent and the colour resembles that of levender.

Leaves are mixed with orange juice to make it more tasty. A delicious drink is prepared by mixing with lemons, pineapple, or orange juice. It is also used in cosmetic industry.

PATCHOULI (*Pogostemon cablin* (Blanco) Beath)

Dr. A.S. Gontia
Principal Scientist,
Deptt. of Plant Physiology, JNKVV, Jabalpur (M.P.)

Botanical Name	:	*Pogostemon cablin (Blanco Beath)*
Family	:	Lamiaceae
Common Name	:	Pathchouli

Introduction

Patchouli belongs to the family Lamiacea of Botanical Kingdom. It is native of Philippines. It grows wild in Malaysia, Indonesia and Singapore. The dry leaves of patchouli on steam distillation yield oil, which in used is perfumery industry.

It blends well with essential oils like vetiver, sandalwood, geranium and lavender. In India it is introduced in 1941 in M.P. The Johore species provides oil of good quality. Other species are Singapore and Indonesia. Its worldwide production is around 700 – 800 tons/year.

Description of Plant

It is an erect, branched, pubescent herb, 0.5 to 1.0 m high, aromatic when crushed. Leaves ovate to oblong ovate, acute or obtuse, simply or doubly crenate serrate, on both surfaces, more or less densely tomentose. Stem swollen on the nodes, spikes terminal and axillary, panicled, dense, 2.5 – 14 cm long calyx 5, corolla lobes obtuse, pink purple or white violet blotched on all segments. Flowering takes place in January to February.

Soil and Climate

Patchouli adapts itself to a wide range of soil and climatic conditions. However, it grows well in partially shaded, well-drained fertile soil with evenly distributed rainfall. A temperature range of 22 - 28^0 C and average humidity of 75% have been found ideal for its growth.

Cultivation

The main field for transplanting should be thoroughly disced and tilled.

Raising the Nursery

Patchouli is mainly propagated vegetatively. Nursery which has to be under shade could be raised at any time during the year. Cuttings are prepared in the morning or evening to minimize desiccation. Cuttings of 10 - 12 cm long having 3 - 4 nodes and leaves with axillary buds at the axils of leaves would be more suitable. The basal end of cuttings should be neatly cut in oblique from just about 1cm below the node. Treating the cuttings with solutions of Indole acetic acid 1000 ppm, Indole butyric acid 500 ppm or Naphthlene acetic acid 1500 ppm will induce early rooting in places of cuttings. The cuttings should be planted in nursery beds or polythene bags with the help of suitable dibbler at a spacing of about 5 cm. Aeration, partial shade and regular watering are essential for early rooting. The cuttings take about 30 –35 days for rooting in nursery.

Transplanting

After about 6 weeks the cuttings are ready for transplanting. Patchouli is shade-loving plant. It can be successfully grown as an intercrop with coconut. Except during heavy rains and hot summer periods, patchouli can be transplanted any time during the year. Transplanting should preferably be taken up during the evening hours with a distance of 90 x 90 cm. in the field. The transplanted seedlings droop initially but recover by the following morning. The plants should be watered every day for 3-4 days and subsequently on alternate days for 10-15 days. By this time the seedlings will have established well and thereafter irrigation once in every 3 days will suffice.

Irrigation

For getting good yield of the crop, the area should receive good and evenly distributed rainfall. Artificial irrigation is, therefore, necessary in areas where rainfall is scanty.

Fertilizer Application

Patchouli requires rich nutrient soils. Normally, a basal dose of 25 kg N, 50 kg P_2O_5 and 50 kg K_2O/ha is given in the form of urea, super phosphate and murate of potash. After about two months for each harvest 50 kg N is applied in two split doses, the first dose just after leaf harvest and the second about two months later. In total, 125 kg N per hectare per year is applied to the crop.

Interculture

The crop many require weeding after about 6 weeks from planting and one hoeing after each harvest.

Harvesting

The first harvest of the crop is taken about 5 months after transplanting. The stage at which the crop has to be harvested is very important for good yield and better oil quality. It has to be harvested when the foliage becomes pale green to light brownish and stand emits characteristic patchouli odour which could be easily smelt particularly in morning hours. Subsequent harvest can be taken after every 3-4 months depending much upon the local conditions and management practices. The crop can be maintained for about 3 years. The first 2 or 3 harvests give good yield and high quality oil, harvesting is done with the help of small sharp shear or secateur. It is essential to leave 4-6 juvenile sprouting buds at the basal region for regeneration.

Drying

The harvested material is spread out under shade in thin layers in order to provide proper aeration. Proper drying is necessary for obtaining maximum yield and good quality oil. Drying normally requires 3-6 days depending upon sunshine and atmospheric humidity.

Oil Extraction

Shade dried leaves of patchouli after removing leaf petiole are subjected to steam distillation for obtaining oil of patchouli. The dried leaves could be immediately distilled or stored for some time as per convenience. Matured leaves contain 5.7% and new or sheded leaves 4.1-5.0% oil contents on oven dry weight basis. Research conducted at Indian Science Institute Banglore indicated 60-65 kg/ha oil productions. Oil contains 40-50%, terpenes. According to Mukarji (1981) Patchouli oil contains Beta alimine 1%, caryophilin 20%, Alphaguine 15%, Alpha Bulnesne 25%, Alpha guine oxide 1%, Alphabulnesne oxide 4%, caryophilin oxide 2%, nar patchouli oil 0.5%, patchoili alcohol 30% and pogostol 1%. Patchouli oil is a viscous liquid, yellowish green to brownish in colour and possess a warm aromatic, spicy fragrance. The characteristic odour in patchouli is due to presence of patchouli alcohol also called patchouli.

Uses

Patchouli oil blends well with sandalwood, geranium, vetiver, ionones, cederwood derivatives, clove oil, lavender, bergamot etc. Oil is almost a perfume by itself. It is widely used in soap, cosmetics, tobacco and incense. The oil gives one to the finest attars when blended with sandalwood oil.

The oil possesses antibacterial activity and used as an ingredient in insect repellant preparation. The oil was found effective against microbacterium tuberculosis. The dried leaves are used for scenting wardrobes. The leaves and tops are added in bath for their ant rheumatic action. It is also used as a masking agent for alcoholic breath.

Storage

Immediately after distillation all free water in the oil should be completely removed and the remaining traces of water can be removed by adding anhydrous sodium sulphate @ 20-30 g/litre. The anhydrous sodium sulphate can be left in the soil for 4-5 hours after which the oil is filtered to remove the sodium sulphate and other sediments, which might come during distillation. The untreated oil develops high acid value, low ester value etc. resulting in to deterioration of the oil. Oil should be stored in aluminum container filled up to the brim and made airtight. The odour of the oil improves on ageing.

Pests and diseases

Root knot – It is caused by nematode, Meloidogyne incognita (Kofoid and white) chitwood. The nematode is an obligate parasite, which enters in a root by thrust with the mouth spear, usually in the region of cell elongation or the base of lateral roots. They reach cortex, multiply and cause galling, which is called as root knot. The diameter of the knot varies from 0.2cm to 5.0 cm. Heavily infected plants show stunting of top growth and wilting symptoms. Generally symptoms become apparent after about 8 months. To control the disease infestation nursery should be raised from healthy mother stock under nematode free conditions. The plot should be treated with carbofuran (furadan) @ 20kg/ha. First dose of the nematicide should be given as a preplanting treatment and the second dose can be given one year after transplanting. In other methods crop rotation and intercropping with citronella or marigold has also been found effective. Solarization technique involves covering of the land with polythene around 15th may when some moisture is left in the soil. This treatment kills the nematodes.

Leaf blight – It is caused by fungus cercospora sp. Mostly when the plants are almost one year old, brown spots are formed near the margin or at the apical region, which enlarge regularly, coalesce, cover the entire lamina and create drying. Two sprays of Dithane M-45 @ 2.5 g/ litre at one month interval have been found effective against disease.

In case of any caterpillar or leaf Webber attack, Endosulphon 0.1% could be added to the dithane spray for control of insect. Methyl Parathion (0.2%) may also be sprayed for control of caterpillar or leaf-Webber.

KALIHARI [*Gloriosa Superba* (L)]

A.B. TIWARI
Scientist
Deptt. of Plant Physiology, JNKVV, Jabalpur (M.P.)

Botanical Name : *Gloriosa Superba* (L)
Family : Liliaceae
English Name : Glory Lily, Tigersclaw
Sanskrit Name : Langali, Visalya
Common Name : Agnishika, Languli, Karihari.

Introduction

Gloriosa superba is a herbaceous perennial climbing plant, growing between 3.5 to 5 meter in length, but usually trained at 1.5 to 2.0 metrs above ground level. The stem is weak with tuberous roots. The leaves are ovate, lanciolate, geenuminate the tips is converted into spirally twisted to serve as tendrils. The flowers are large, solitary corymbose inflorescence having yellow colour but changing to yellow-red and deep scarlet. Ovary tricarpellary and fruit is capsular having numerous seeds.

Distribution

G. Superba is a native of tropical Asia and Africa, It is found growing throughout tropical India. From N.W. Himalayas to Assam and Karnataka also growing along the Western Ghats. There are other four species are also found in India. *G. superba* also found growing in

Madagascar, Sri Lanka, and Indonesia. Growing area about 2000 ha in India.

Climate

Kalihari is a tropical plant and comes up well in warm humid regions. It can be grown up to 600 m from sea level with a annual rain fall about 373 cm well distributed throughout the year. The temperature for its growth 15-20°C during the day and 10-15°C at night.

Soil

Glory lily prefers sandy loam red or black loamy soil having pH range between 6-7 has been found suitable to raise this crop. It can be grown in sloppy and undulated forestlands. Other soils with medium water holding capacity and good drainage also suitable to grow the crop.

Cultivation

Field preparation

The field should be ploughed until it is brought to a fine tilth. The field must be leveled properly and drainage arrangement made to avoid water logging during rainy season. The field is then divided into subplots to convenient sizes.

Fertilizers

About 15-20t/ha of FYM or compost should be applied at time of field preparation. Chemical fertilizer NPK (50:50:70) per hectare should be applied at the time of plantation and N-60kg/ha should also top dressed in two split applications at 5-8 week after plantation. *Gloriosa* insures vigorous plant, stronger tubers and better flowering to manuring and fertilizers.

Propagation

Kalihari is commercially propagated by v-shaped rhizomes or sexually propagated by seeds. The plants raised from seeds take nearly three to four years to flower. Hence seed propagation is not favoured by the growers.

Gloriosa produce undergrount biforked or V=shaped tuber during the growing season and each of these forks has only one growing bud. Tubers should be handled carefully as they brittle and liable to break easily. If the growing bud is subjected to any kind of damage, the tuber will fail to sprout.

Varieties

Glorisa superba is a newly domesticated crop and there are no identified varieties. Only 10 species of Kalihari all over the world.

Sowing period

The planting is done during June-July under rainfed condition. While under irrigated conditions it is planted in the month of May.

The crop raised by seed, the sowing of seed is done in nursery during May-June and the seedling remain in the nursery for 2-3 year till the rhizome weight should be 40-50 grams at the time of plantation.

Plantation method

One-foot deep furrows are made in May-June in a well-prepared field at spacing on 50-60 cm. The tubers are planted at a depth of 6-8 cm. Keeping a plant-to-plant distance 40-50 cm. depending upon the type of soil.

The planting tuber should not weight less then 50 gms. The plants raised from smaller tubers do not prodce flowers during first year. The large tubers can be divided in to two by breaking them in the middled. For the better crop only healthy tubers should be selected for planting.

Irrigation

After planting of tubers field should be irrigated. After rainy season irrigation at an interval of 20-30 days should be done after September upto maturity of the fruits. Excess watering is harmful to the plants and causes yellow or brown coloured pateles on the leaves, which fall off at early stage.

Stacking

The provision of some kind of support is necessary for better growing of Kalihari. The plant not remain erect due to its bery tender weak stems, hence the bamboos or branches of kher trees should be used for stacking purpose when the plants are about 30 to 40 cm tall. The leaf tips tendril help to climb the plant on support.

Weeding

The weed should be controlled manually, 2-3 weeding are required. During weeding almost care should be taken to avoid any damage to the growing tip, once damaged it does not sprout again during the season.

Plant protection

Insects: Lily caterpiller and green caterpiller controlled by spraying of 0.2% endosulfan or metacid at fortnightly intervals.

Diseases: For the control of leaf blight spraying 0.3% Dithane M-45 at weekly intervals.

Rhizome rot can be controlled by drenching the soil with Bavistin at 0.2% at the time of plantation of tuber.

Harvesting

The fruits are harvested after 170-180 days after transplanting of the tuber in the month of November - December. Flowering started after 55 days and continues to flower and fruit till October. The fruit takes 110 to 120 days to reach maturity. The Right stage of fruit harvest is when this capsule starts turning light green from dark green and the skin of the fruit shows a shrunken appearance and becomes light weight.

After picking the capsules dried in shade for 10-15 days. The seeds and pericarop (Fruit wall) are separated manually and dried for a week in the shade.

The tubers are harvested after five years of plantation, cut in to small pieces and dried in shade, stored in bags.

Yield

The yields of seeds differ greatly, depending upon the vigour and age of the plant. The yield in the initial stage will be low, but it gradually increases in the subsequent years. After three years, from a well-managed field seed yield about 200 to 225 kg/ha. The yield of pericarp (Fruit wall or Husk) is 150-160 kg/ha annually.

The tubers roots yield comes 25 to 30 q/ha after five years of the plantation.

Chemical constituents

The flowers, leaves and rhizomes contain colchicines, N-formyldeacethyl colchicines demethylcolchicine and lumicolchicine. Rhizome also contain gloriocine. Leaves Chelidonic acid-2 hydroxy-6methoxybenzoic acod and B-sitosterol glucoside. Colchicine demethyl colchicines and colchicoside has been reported from seeds.

Uses

Rhizomes-Tonic, anti cancer, anti malarial, stomach-ache, purgative, cholagogue, anthelmintic, alterative, feverifuge, andti leprotic, antidote for snake bite and scorpion sting, abortifacient, in piles, in colic, parasitical affection of skin, glossitis of cows, starch from rhizomes antigonorrhoeic leaf antiasthmetic. Rhizomes-Oxitocic, fresh juice of plant-leaf stimulant, root antibiotic.

ISABGOL (*Plantago ovata* Forsk)

R.S. PATEL
Associate Professor
Dept. of Plant Physiology, JNKVV Jabalpur

Botanical Name : *Plantago ovata*

Family : Plantaginaceae

Common Name : Isabgol

Distribution of area and production

Isabgol crop has acquired the place of the 'dollar earner' crop of north Gujarat, where it is being cultivated on about 50,000 ha. With a production of 40,000 tonnes, southern Rajasthan, Punjab, Maharastra and Uttar Pradesh states are also grown Isabgol crops. About 75-80% of the total annual produce of North Gujarat is exported.

Identifications of plant

Isabgol is a stem less or short stemmed, highly cross pollinated rabi crop. Plant height about 30-40 cm. Alternate leaves, which covers the stem. Stems and leaves with fine hairs. The flowers are white minute, four parted. The capsule is ovate 8.0 mm long. Seeds are ovate and 1.8 to 3.8 mm long brown gray colour. Seeds are covered with a translucent membrane, known as husk.

Varieties

1. Gujarat Isabgol - 1 (GL - 1) Released by Gujarat Ag. University.
2. Gujarat Isabgol - 2 (GL -2) Released by Gujarat Ag. University.
3. Niharika Released by CIMAP Lucknow. as a high yielding variety.

Soil

Light soil is best suitable for Isabgol. Silty loams with 5.00 to 7.00 pH and low nitrogen and moisture content soils are gave better yield.

Climate

Isabgol require warm temperate regions, cool dry weather and sown winter (Rabi) season. First week of November sown crop gave higher yield. Better seed germination on 20^0 C temperature.

Cultivation

a. The field must be free from weeds and make fine tilth for good germination. Mix 10-15 tonnes FYM/ha at the time of field preparation.

b. Seeds and sowing

(i) Fresh and bold seeds are selected for sowing.
(ii) Seed rate 6-8 Kg/ha. seeds
(iii) Method of sowing
  (a) Broad cost method
  (b) Line sowing method

An important thing is that Isabgol seeds are small and light, hence the seeds are mixed with fine sand or sieved FYM. After that the seeds are sown broad cost or line sown. Seed must be shallow cover with fine soil or sieved FYM otherwise the germination are affected if seeds are in deep or above the soil germination are not proper and uniformly.

Irrigation

After sowing light irrigation through sprinkler irrigation system is recommended because if flood irrigation is given the seeds are disrobed their placement.

Fertilizers

Isabgol require following doses of fertilizers.

1. FYM - 10 – 15 Tonnes per hectare at the time of field preparation.
2. Nitrogen - 50 Kg/ha
3. $P_2 0_5$ - 25 kg/ha
4. $K_2 0$ - 30 kg/ha

Out of 50 kg Nitrogen, 25 Kg nitrogen at the sowing, full dose of $P_2 0_5$ and K_2O are applied at the time of sowing. Half dose (25 Kg/ha) is applied after one month sowing.

Irrigation

First irrigation after one month of sowing. Second irrigation after flowering stage. Third irrigation is given if necessary or in light soil.

Inter culture operation

The first weeding is done after about 20-25 days of sowing. Second weeding after 40-45 days of sowing if required.

Insect pests and diseases control measure

Some times white grubs and termites are damaged to crop at root zone. These can be controlled by broadcosting of phorate 10 G, at the rate of 10 kg/ha at the time of field preparation.

Aphids also damage the crops, aphids are controlled by spraying 0.3% dimethoate.

Downy mildew diseases

This disease appears at the time of spike initiation. The diseases intensity will be very high in humid weather condition.

Control measures

1. Boadeaux Mixture (6: 3: 100) at the rate of 600 to 800 litres per hectare.

2. Spraying of copper oxychloride, dithane M-45, Dithane Z-78 or any other copper fungicide at the rate of 2.0 to 2.5 g/litre water.

Harvesting and threshing

On crop age of 110-130 days the leaves colour change green to yellowish, spike colour turn to brown. Means Isabgol crop ready for harvesting. Harvesting done on morning time so that shetering are minimize. Harvested crops are dried for 2 or 3 days. Dried crops are threshed out by tractors or beaten by stick. Threshed seeds are dried and store in bags at dry place.

Yield

Variety	- Yield/kg per hectare
GL-1	- 800-900 kg.
GL-2	- 1000 kg.
Niharika	- 1000 – 1100 kg.
Ratio of husk and seed	- 25: 75

Chemical contents

The Isabgol husk is used for its mucilage content mucin or mucilage are generally acidic mucoply-saccharides. They are usually attached to proteins to form proteoglycans, in which the polysacchrides make up most of the weight. Normally mucin consists of recurring disaccharide units each of which contains a derivative of an aminohexose, usually D glycosamine or D galactosamine. The polysaccharides isolated from mucin are slippery and lubricating proteoglycons of mucous. These are jelly like ground substance or intercellular cement.

Uses

(A) Seed -

1. Sowing purpose.
2. The seed has also cooling and demulscent effects and is used in Ayurvedic, Unani, and Allopathic system of medicine.

Contains – Oil and small amounts of glycoside a cubin and tannin.

(B) Seed Husk

It is economic parts contains a colloidal mucilage 30%.

(i) Xylose
(ii) Arabinose
(iii) Galachironic
(iv) Rhannose
(v) Galactose

Contains

(a) Antidiarrhoeal drug.
(b) Beneficial in chronic dysenteries of amoebic bacillary origin.
(c) Used for treating constipation and intestinal disorders.
(d) Seeds and husks are used to cure inflammations of the mucous membrances of gastro-intestinal and genito urinary tracts, duodenal ulcers, gonorrhoea and piles.
(e) It can also be used as a cerrical dilator for the termination of prengnancy.
(f) It has a place in dyeing, calico printing, in the ice-cream industry as a stabilizer, also in confectionery and cosmetic industries.
(g) The seed without the husk, which contains about 17-19% protein is used as cattle feed for increasing the milk

STEVIA SWEET HERB (*Stevia rebaudiana*)

V.K. AGRAWAL
Professor and Head
Deptt. of Plant and Physiology, JNKVV, Jabalpur (M.P.)

Botanical Name : *Stevia rebaudiana*

Family : Asteraceae

Common Name : Meethi Tulsi

Description

Stevia rebaudiana family - Asteraceae, is a perennial leafy green plant commonly known as sweet herb of Paraguay. Stevia ia a small shrubby perennial grqing up to 65 to 75 cm tall with sessile, oppositely arranged lanceolate to lblanceolate leaves serrated above the middle flowers are small, white and arranged in an irregular cyme, Stevia is short day plant with critical length of 13 hrs. The plant is commercially grown in china, Japan, south Korea, India, Ukraine, Australia, Canada, Mexico, United States of America, UK, Belgium, Germany, Spain and many countries in South America.

- The plant is gradually gaining interest all over the world due to presence of zero calorie sweetener compounds such as Stevioside (8.0-11.0%) and Rebaudioside A (2-4%). Stevia holds 47% of the sweeteners market in Japan. Stevia is lesser known plant and accepted only as a dietary supplement in many countries including USA.
- The market opportunity appears great. Statistics indicate that in some countries up to 30% of their needed sugar is replaced by chemical-based (synthetic) stevioside-like sweetness products.

Benefits

- No calories and it is natural
- pH and heat stable to 200 °C (392 °F)
- Does not affect blood sugar levels
- Non-toxic and non-addictive sweetener
- Non-fermentative and non-discolouring
- Bactericidal activity
- Potent: 250 to 300 times sweeter than sugar
- Various medicinal uses

Uses

In Japan alone, an estimated 50 tons of Stevioside is used annually. The market opportunity appears great. Statistics indicate that in some countries up to 30% of their needed sugar is replaced by chemical-based (synthetic) stevioside-like sweetness products. Following are the main application areas:

- Flavour enhancer
- Herbal tea
- Preparation of Pharmaceutical products
- Food and beverages application as a replacement for sugar and artificial sweeteners and also combined with other sweeteners
- Added to products such as chewing gum, toothpaste, mouthwash and even an anti-smoking lozenge.

Medicinal Uses

Hypoglycemic action : Helpful for hypoglycemia and diabetes.

Cardiovascular action : The long term use would have a cardio tonic action.

Antimicrobial action : Inhibits the growth and reproduction of bacteria and other infectious organisms.

Digestive Tonic action : Improves digestion, and overall gastrointestinal functions (Kinghorn, D.,1991)

Skin Problems : Effective when applied to acne, seborrhea, dermatitis, eczema.

Types of Stevia Products

1. *Stevia leaves*

Fresh leaves	Dried Leaves
Tea Cut Leaves	Ground leaves (powder)

2. *Liquid Extracts*

Dark or Clear: dissolved in water or alcohol or glycerin.

3. *Stevia Extracts*

40-50% Sweet Glycosides
85-95% : greater concentration

4. *Stevioside*

Stevioside is the purified or most highly processed form of Stevia.

5. *Stevia Blends or spoonable Stevia*

Fillers used: Dextrose, lactose, Maltodextrin, Fructo-oligosaccharides

6. *Stevia Packets*

Cinvenient and pre-measured servings.

7. *Stevia quick dissolving tablets*

Mainly used to sweeten beverages.

Climate and Soils

Stevia is grown as a perennial in subtropical and mild temperate regions but must be grown as an annual in mid and high altitudes. This is a short day plant. The concentration of steviosides in the leaves increases when grown under long day conditions. Average maximum and minimum temperature should not exceed 10^0-37^0 C during the active crop growth period; the crop thrives well in relative humidity ranging from 65 to 85 %. Stevia can toerate rains but is prone to frost. The Stevia crop has a capacity to regenerate again in spring even after burning of the aerial parts due to frost injury. Stevia can grow well on a wide range of soils. It performs beter in well-drained red soil and sandy loam soil. The soil should be well prepared having good textured soil to support plant growth. Optimum supply of moisture and drainage is necessary for proper growth of the plant. Soil pH should be within the range of 5.5-7.5. Most of the soils good for vegetable are also suitable for stevia cultivation.

Propagation

Stevia plants can be propagated from cuttings as well as through seeds. In many of the cultivars, seed germination rate is poor and crop raised through seedlings take more time to establish. Therefore, it is best grown through transplanting as an annual or perennial crop.

Nursery Raising

1. Nursery raising through seeds should be done in the month of February-March. Germination of seeds takes place within 10-15 days. Seeds can be sown in plastic trays, pots, wooden boxes or in raised nursery beds of size 1.25 m x 10 m having a good mixture of sand, soil and organic manure. Nursery should be irrigated with sprinkler daily in morning and evening for first five days and once a day during next five days, One month old seedlings of 5-7 leaves stage and 8-10 cm height become ready for transplanting.

2. Saplings can be raised successfully by vegetative means using terminal cuttings of 10-15 cm height with 3-6 nodes. Stem cuttings should be taken from actively growing plants during the months of mild weather avoiding extreme cold or hot months. Lower leaves are trimmed to facilitate planting. Cuttings are raised in pots or poly sleeves. In nursery beds the cuttings are planted at 15 x 15 cm spacing keeping one node inside the soil. Cutting raised under partial shade and under high humidity develops roots at faster rate within 8-10 days. Later on irrigation is done twice a

week or as and when required. Spraying of Bavistin (0.1%) and Thiodan (0.25%) is done to keep the plants free from diseases and insect pest.

Transplanting in field

Rooted cuttings of 4-5 weeks old are transplanted in the field. Stevia is generally transplanted in the field during March-April and June-July avoiding periods of extreme climate. Land should be prepared by repeatedly ploughing and harrowing. Irrigation and drainage channels should be laid out as per the layout of the field. Transplanting is done in furrows at a spacing of 45 x 45 cm.

Manures and fertilizers

Most like leafy vegetables, Stevia is fertilized through organic manures considering a low nutrients consuming plant. A well-rotted 25-30 ton/ha farmyard manure application at the time of field preparation is sufficient to grow a successful crop.

Irrigation and drainage

Stevia cannot tolerate drought, therefore frequent irrigation is required, First irrigation is given immediate after transplanting and another after 2-3 days. Subsequent watering is done weekly till onset of monsoon. However, actual numbers of irrigations are based on moisture holding capacity of the soil and occurrence of the natural precipitation.

Weeding and hoeing

Removal of weeds can be done manually. Hence, two-time hand weeding and one hoeing are sufficient to keep the weeds under control.

Harvesting

Harvesting is done manually leaving 8-10 cm stem height from the ground level. A first harvest is taken 75-90 days after transplanting in the month of June-July. Subsequent second harvest is taken after 60-75 days of the first harvest in early September at the time of flower bud initiation. In case of late transplanting crop grown for single cut, harvesting is done after 3-4 months of transplanting and continues till flowering begins as the sweetener is maximum in the leaves till the plant flowers. Perennial crop may continue up to four years, once planted in the same field. Maximum amount of leaves are produced in 3rd or 4th years. Flowering of the plant should be avoided and pinching of the apical bud should be done to enhance busy growth of the plant with side branches.

Yield

Average fresh biomass yield of 25-30 ton/ha/yr out of two harvest is obtained which give dry herb weight of 4-5 ton. Leaf powder being commercial part has a significant meaning in terms of its yield. Dried leaf yield of 13-16 q/ha can be separated from the total biomass yield.

Post harvest practices

After harvesting, the whole plant is dried & the leaves are separated from the stems for further processing. The stems have very low concentrations of sweet glycosides and are removed to minimize processing cost. The dried leaves are powdered and stored in airtight containers or plastic bags in a cool place. Dried leaf powder can be directly marketed without further processing by the farmers. Water extraction of the dried leaves is followed by clarification and crystallization processes which facilitate production of stevioside and rebaudioside. Most commercial processes consist of water extraction, discoloration and purification using ion-exchange resins, electrolytic techniques, or precipitating agents. Most of the commercial processes are protected by way of patent and other IPR protection ways.

Chemistry of *Stevia rebaudiana*

There has been interest in the chemistry of the sweet principles dates back from early in the century but significant progress towards chemical characterization was not made until 1931, with the isolation of Stevioside. The sweetness is due to the presence of eight diterpene glycosides based on the kaurene skeleton. They are stevioside (8-12%), Rebaudioside A to E, Dulcoside A & steviolbioside. Stevioside and Rebaudioside A, in pure state, are 300 and 450 times sweeter than sugar.

Extraction and refining

There are hundreds of patents for stevia extraction processes on worldwide basis and Japan alone has over 150. The processing steps involved in the production of sweeteners are extraction, pretreatment, separation and refining.

Extraction with water is the most effective solvent. The pH, temperature, mode of extraction and solvent type affects the impurities (protein, pigment, pectin's and flavonoids) in the extract. Selection of proper extraction methods to minimize the impurities from the crude extract affects the product cost significantly.

Pretreatment step reduces the impurities by precipitation with inorganic salts such as Al_2SO_3, $Ca(OH)_2$ at pH 8-11. This can also be done by a membrane process. Micro filtration provides chemical free pretreatment and remove pigment impurities absorbing at 420 and 670 nm. Other polymeric membranes reported to remove pigments more effectively with about 90% stevioside recovery.

Separation and refining

Separation and Refining steps include usage of polymeric absorbent resins, ion exchange resins or Ultrafiltration membranes. The flow rate, pH and eluents greatly affect both the separation efficiency and treatment capacity of the columns. Overall stevioside purity may be achieved to 80 to 85% by the above methods. Still higher pure stevioside and Rebaudioside-A are achieved by recrystallization in alcohol and water mixtures.

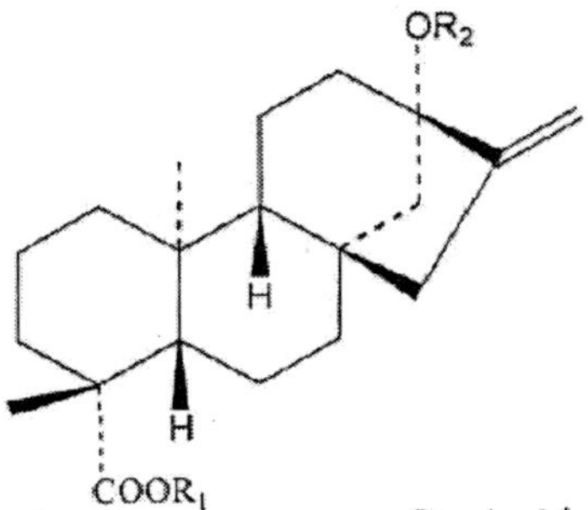

R_1- glu and R_2- glu-glu — Stevioside

R_1- glu and R_2- glu-glu-glu — Rebaudioside A

List of Indian Buyers of Stevia Herb/Products

– Jayalakshmi industries
  Buyers of stevia leaves, leaf powder.
  3 A, E and E Industrial Estate, Hosur –635 109, India
  Phone: +(91)-(4344)-544672 Fax : +(91)-(4344)-43365

– Saroj Agro
  Buyers of stevia plants and products
  Shambika Path, New Dilian, Sasarm – 821 307, India
  Phone: +(91)-(6184)-250320 Fax : +(91)-(6184)-250320

– Herbal, Inc.
  Buyers of dried stevia leaves.
  Raghu Rama Street, Sri Laxi Residency,
  GG-I, Vijayawada – 520 010, India
  Phone: +(91)-(866)-2486152

– Sri herbs
  Buyer of dried stevia leaves
  Raghr R1ama street
  Sri Laxmi Residency
  GF-I, Vijaywada 52010
  Phone : +(91) – (866) - 248615

Cost of cultivation of Stevia rebaundiana

Total Expenses for 4 Years/ha	=	4,17,870.00
Gross Returns for 4 years/ha	=	6,00,000.00
Net Returns for 4 years/ha	=	1,82,130.00
Net Returns/ha/year	=	45,533.00

GHRITA KUMARI (*Aloe vera*)

V.K. AGRAWAL
Professor and Head
Deptt. of Plant and Physiology, JNKVV, Jabalpur (M.P.)

Botanical Name	:	*Aloe vera Mill,* syn, *A. Aloe barbadensis*
Family	:	Lilaceae
Vernacular Names	:	Ghrita kumari or Guar Patha.

Aloe vera linne or Aloe barbadensis Miller is a succulent of Aloe family with its origin in the African continent. Its thick leaves contain the water supply for the plant to survive long periods of drought. These leaves have a high capacity of retaining the water even in very warm and dry climates and therefore this plant can survive very harsh circumstances where most other vegetation disappears. When a leaf is cut, a orange-yellow sap drips from the open end. As a drink this bitter sap has a very strong laxative effect. A "stabilized" product is incorporated in a wide variety of preparations, including juices, gels, ointments, creams, lotion and shampoos.

Distribution

The genus Aloe is widespread throughout the entire African continent, but the tropical regions are particularly rich. It is also grown as an ornamental in India. *Aloe Vera* is cultivated in fairly large areas many parts of India viz. Chhattisgarh, Maharashtra, Madhya Pradesh and Gujrat.

Botany

It is a perennial herb with short, stout and thick stem. Leaves are sessile, crowded, lanceolate, pale green, fleshy, margins spiny. Scape is longr than the leaves. Flowers are in dense racemes, perianth cylindric; pendulous and yellow in colour.

Chemical constituents

All Aloe species contain anthraquinone glycosides. Barbaloin (aloe-emodin anthrone C-10 glucoside) is the major active constituent. Aloes also contain isobarbalin, aloe-emodin, resins, aloetic acid, homonataloin aloe-sone, chrysophanic acid, chrysamminic acid, galactouronic acid, choline choline- salicylate, saponins, mucopolysacchrides, glucosamines, hexuranic acid, coniferyl alchohol.

Useful parts

Leaves

Leaves exude a bitter liquid, which is dried and known as "bitter aloes." They also contain a clear gel, which is a soothing skin remedy.

Cultivation

Climatic requirements

Flourishes in a variety of climate and can withstand drought.

Soils

Driest and poorest soils are good for economic yield of aloe. It can be grown on a variety of soils. However, water logged soils are not suitable for its cultivation.

Propagation

Aloe vera is propagated by breaking off small rooted plantlets, through seed, sucker and cutting. A total of 36,500 suckers are required for one-hectare plantation.

Nutrients

No chemical fertilizer is recommended for its cultivation. It thrive well when grown with or without FYM. In India,

Aloe is raised as organically and only farmyard manure is applied @ 12-15 tones/ha. In standing crop cow dung solution is applied for nutrient and peat management.

Pest Infestation

No major insects or disease have been reported in India.

Maturity, harvesting and yield

The plant takes approximately four years to reach economic maturity and has a lifespan of about 12 years. Crop is ready to harvest about 8 months after sowing India, the average yield for organically grown Aloe is about 12 tones/ha (on fresh weight basis leaves). To collect the gel and bitter liquid, the leaves are cut and drained as required. The plants can be harvested every 6-8 weeks by removing 3 to 4 leaves per plant.

***Aloe vera* helps beauty treatment : Aloe vera has a long history as a skin lotion.**

***Aloe vera* helps in healing of wounds :** The bulk of the *Aloe vera* leaf is filled with gel, 96% water with other 4% containing 75 known substances. Applied to wounds, aloe vera gel is a mild anesthetic, relieving itching, swelling and pain: it also has antibacterial and antifungal properties, increases blood flow to wounded areas and stimulates fibroblasts, the skin cells responsible to wound healing.

***Aloe vera* gel soothes burns:** The burns healed more quickly when aloe gel is applied on mild burn.

***Aloe vera* minimizes frostbite damage:** Aloe gel is help full in frostbite. It prevented a decrease of blood flow to the frozen tissues, a common cause of tissue loss in frostbite.

***Aloe vera* screens out radiation:** Aloe vera gel protects against skin damage from X rays,. It is an effective antioxidant, mopping up the free radicals caused by radiation and that it protected healing substances, super oxide dismutase (an antioxidant enzyme) and glutathione (an amino acid which stimulates the immune system).

***Aloe vera* protects form lung cancer:** Aloe vera juice prevents human pulmonary carcinogenesis (lung cancer) and is preventive or suppressive against various human cancers.

***Aloe vera* eases intestinal problems:** Aloe vera juice can be effective for treating inflammatory bowel disease.

***Aloe vera* reduces blood sugar in diabetes:** Aloe vera juice reduced the blood sugar levels in diabetics. It can reduce blood sugar levels by an average of 45%, with no change in their total weight.

Aloe vera: its potential use in wound healing and disease control in oral conditions

Aloe vera has been shown to enhance defense mechanisms, and it has a variety of components to help combat periodontal disease and other oral conditions. There are eight main uses of Aloe vera in dental practice:

1. Applications directly to the sites of periodontal surgery.
2. Applications to the gum tissues when they have been traumatized or scratched by toothbrush dentifrice abrasion, sharp foods dentifrice abrasion, sharp foods, dental floss and toothpick injuries.
3. Chemical burns are relieved quickly from accidents with aspirin.
4. Extraction sites respond more comfortably and dry sockets do not develop when aloe vera is applied.
5. Acute mouth lesions are improves by direct application on herpetic viral lesions, aphthous ulcers, and canker sores and cracks occurring at the corners of our lips. Gum abscesses are soothed by the applications as well.
6. Other oral diseases chronic in nature respond with Lichen Planus and Benign Pemphigus. Even gum problems associated with AIDS and leukemia patients

receive relief. Migratory glossititis, geographic tongue and Burning Mount Syndrome are improved.

7. Denture patients with sore ridges and ill-fitting dentures and partials can benefit as fungises and bacterial contamination reduce the inflammatory irritations.
8. Aloe vera can also be used around dental implants to control inflammation from bacteria contamination.

Doses

For constipation, a single 50-200 mg capsule of aloe latex can be taken each day for a maximum of ten days, Tropically for minor burns, the stabilized aloe gel is applied to the affected area of skin three to five times per day; Treatment of more serious burns should only be done after first consulting a health care professional. For internal use of aloe gel, 30 ml three times per day is used by some people.

Side effect

Except in the rare person who is allergic to aloe, topical application of the gel is harmless.

Products and applications

Gel

Fresh: Apply the split leaf directly to burns, wounds, dry skin, fungal infections and insect bites. Take up to 2 tsp in a glass of water or fruit juice, three times a day, as a tonic.

Ointment: Split Aloe vera leave to collect a large quantity of gel and boil it down to thick paste. Store in clean jars in a cool place and use like the fresh leaves.

Tonic wine: Fermented aloe gel with honey and spaces is known as kumaryasav in India and is used as a tonic for anemia, poor digestive function and liver disorders.

Inhalation: Use the gel in steam inhalant for bronchial congestion.

Leaves

Tincture: Use 1-3 ml per dose as an appetite stimulant or for constipation. The taste is unpleasant.

Powder: Use 100-500 mg per dose or in capsules as a purgative for stubborn constipation and to stimulate bile flow.

ALOE VERA PROCESSING
PROCESS FLOW SHEET
PROCESS FLOW

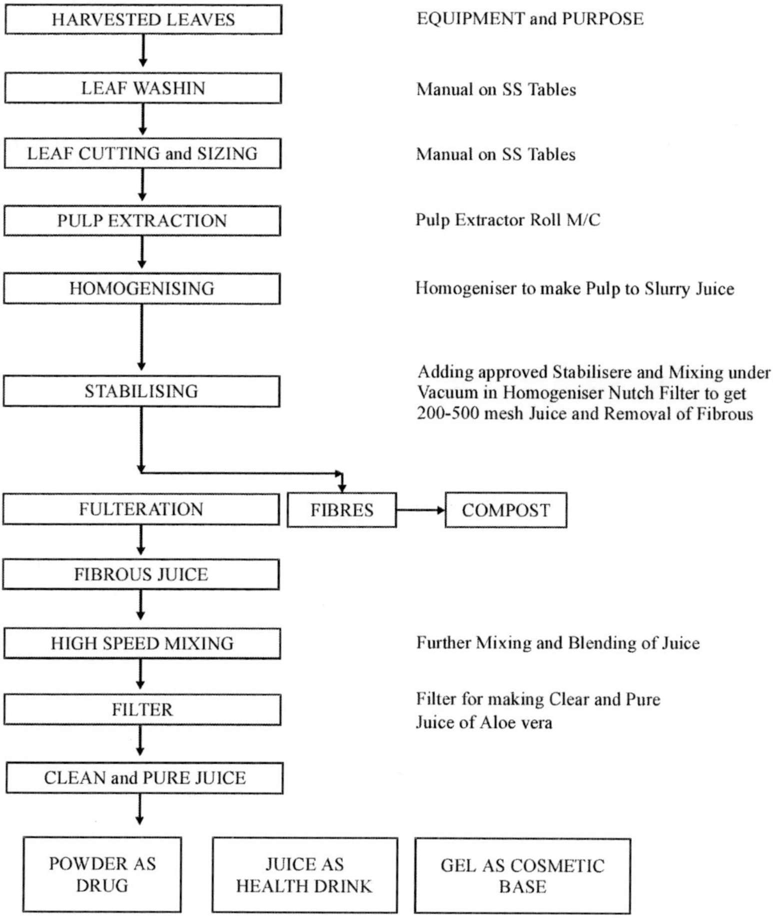

LEMON GRASS (Cymbopogon *flexuosus* Nees ex. Steud Wats.)

S.D. UPADHYAYA
Professor
Department of Plant Physiology, J.N.K.V.V. , Jabalpur

Botanical Name	:	*Cymbopogon flexuosus*
Family	:	Poaceae (Graminae)
Vernacular Name	:	Hindi: Lemon grass, Gandha-trina; Malayalam and Telugu: Vasana-pillu, Sanskrit: Bustrina, Tamil: Nimma-gaddi, Kannada: Nimbehullu

Introduction

Lemon grass (*Cymbopogon flexuosus* Nees ex. Steud Wats) is the source of lemon grass oil obtained from the leaves and shoots of the plant. Three types of lemon grass are cultivated in India as the important source of Citral.

(I) East Indian lemon grass - *C. flexuosus*
(ii) West Indian lemon grass - *C. citratus*
(iii) Jammu Lemon Grass - *C. pendulus*

All these are perennial grasses and grow upto 2.5 m height. *Cymbopogon flexuosus* is cultivated mostly in Kerala as "Kochin oil" in trade. The annual production is 1000 M. T. per year and the world demand is higher. The annual export of lemongrass oil is about Rs. 5 Crores. Tropical and subtropical conditions are suitable.

Distribution

The species is considered to have originated in India. It grows wild in many parts of the world i.e. tropical and subtropical parts of Asia, Africa, America, Bangladesh, China, Madagascar, Thailand, and West India Islands.

Description of the plant

C. flexuosus grows to a height of about 3 m. The leaves of the plant are linear, lanceolate, 125 cm long and 1.7 cm broad. The panicles are very large, drooping, lax, grayish or grayish-green, rarely purple-tinged, with the raceme pairs in dense masses. The plant is spreading, 100-135 cm tall, slightly hairy, the lower glumes of the sessile spikeletes are 3-4 and rarely 4-5 mm long, 1 mm wide, with 1-3 definite or obscure introcrainal nerves, shallowly concave, with one or two depressions.

Climate

Warm and humid climate is ideal with uniform distribution of rainfall (250-300 cm per annum). It can be grown under unirrigated conditions as rainfed crop.

Soil

Well-drained sandy loam soils are better. However, it can be grown in variety of soils having pH range of 4.5 to 7.5. It doesn't tolerate water logging. However, it can be grown in slopes of hilly forestlands and wastelands.

Variety

Odakkali O.D.19, S.D.68 which yield 50 t/ha herbage and 80 to 85 kg oil/ha recovery is 1.2 to 1.5 % of oil. RRL-16 is developed from Jammu. A Hybrid CKP-25 is also developed from RRL Jammu, which has higher oil yielding capacity by 50% and 140% over OD19 and RRL-16, respectively.

Pragati, Praman, GRL-1, Cauvery, Krishna, and Chirharit from CIMAP Lucknow are released which have higher yields.

Propagation

It can be propagated by seed as well as by slips (rooted tillers).

Seed rate

4 to 5 kg/seed is required for raising the nursery to plant one hectare are as per recommendation of CIMAP while ICAR has recommended 10 kg seed. Direct seeding can also be done with a seed rate 15 kg/ha.

Nursery raising

Nursery beds are prepared and sowing is done in the month of April-May. The seeds germinate within 7 days. The seedlings are maintained in nursery up to 60 days, and then transplanting is done in the main field.

Transplanting

The slips are transplanted at a distance of 50 cm x 40 cm while seedling raised in the nursery are transplanted at a distance of 30 cm x 30 cm. The slips are planted 5 to 8 cm deep. The appropriate time for transplanting is first week of July while under irrigated condition it can be transplanted at any time.

Manures and fertilizers

This crop can be grown without manures and fertilizers, however the yield will be higher under fertilizer application. 20 t FYM/ha at the time of field preparation should be applied in the poor soils. A fertilizer dose of 75:30:30 kg NPK/ha is recommended. Half of the N and full dose of P and K is applied at planting while half N is applied in two-split dose after first and second cuttings under irrigated conditions.

Weeding

One weeding is done 20 days after transplanting later on weeding is not required as the crop smothers the weeds.

Irrigation

This can be grown under unirrigated conditions but for getting higher yields it should be irrigated after each cuttings.

Harvesting

First cutting is done at 3 months after planting from 20 cm of the ground. In a year, 4 cuttings can be done depending on the irrigation and fertilizer application. Once the crop is planted it is harvested upto five years.

Yield

The oil is obtained from the green herbage. The yield of green herbage from 4 cuttings is ranged from 20 to 25 tones/ha, which yield 60 to 80 kg oil on distillation. Based on fresh weight of the leaves the oil percent is about 0.3-0.5%.

Distillation

The water-steam distillation done for extraction of essential oil from the leaves. The major source of loss is by oxidation and resinification of the essential oils. So, if the material is to be stored before processing, it should be kept in a dry atmosphere with limited air circulation. The cut grass, when stored in the shade, can increase the oil recovery up to 96 hours and storage for a further period will only decrease the oil yield. The essential oils are enclosed in the oil glands, oil-sacks and glandular hairs of the plant. Therefore, before distillation, the plant material must be cut into small pieces to enable them to directly expose as many oil glands as practically possible. Once the plant material has been reduced in size, it must be distilled immediately. Otherwise, the essential oil, being volatile, will be lost by aerial evaporation. Dipping the chopped lemon grass in sodium chloride solution for 24 hours at 1-2% concentration before distillation has been found to increase the citral content.

Utilization of spent material

The residue thrown out after the extraction of the oil is called spent grass. Cattle relish it when it is hot or is converted into silage by adding a dilute solution of molasses.

It is also an excellent source of manure. It is applied either after composting or in the form of ash by burning. The waste lemon grass is also used as fuel for distillation after drying and as cheap packing material for glassware or other fragile objects.

Economics

The cost of cultivation including distillation of the grass is about Rs. 16,000/ha. The oil production about 80 kg/ha which is sold at the rate Rs. 250 kg and the gross return of Rs. 20,000/- ha is obtain during first year. The net profit of Rs. 4000 can be expected during first year. Subsequently, the net profit will be Rs. 10,000/ha, as the cost of the plants will be reduced.

Chemical constituent

Main constituents of the oil are Citral-a, Citral b and myrcene. Other contents are a and b Pinene, Car-3-ene, dipentene, B-phellandrene, P-cymene, Citronellal, b-elemene, B caryophyllene, citronellyl acetate, geranyl acetate, geraniol, elemol and Linalool.

Use

The oil is used in many pharmaceutical preparations such as pain balm and disinfactants carminative, anticholerin. Leaf: Sudorific stimulent and anticatarrhal. Lemongrass oil is basic raw material for manufacture of many aromatic chemicals, which are used in soap, perfumery and cosmetic industries. Writing and printing papers are prepared by spent grass.

Flatulence: It is given in dose of 3 to 6 drops with sugar as an imulsion for relieving flatulence.

Digestive disorders: It is strengthen the function of stomach and promotes its action. Oil is useful in indigestion, spasmodic affections of the bowels, irritability and cholera.

Fever: Grass induces perspiration brings down the temperature. Raw juice or decoction of the grass may be taken under such situations.

Menstrual disorder: An infusion of the grass mixed with black pepper is given in painful and difficult menstruation. Raw juice or decoction of leaves may taken in such a conditions.

Rheumatism and other joint pains: The grass is used locally over rheumatic joints lumbago and sprains. Lemon grass oil mixed with twice of its bulk with coconut oil is a stimulant ointment for rheumatism lumbago neuralgia, sprains and other painful affections. In chronic cases undiluted oil may be used for better results. It can be also taken internally as for fever.

Ringworm: Leaves of lemon grass are useful in treating ringworm as local application. A paste of the leaves made in buttermilk should be applied on the affected part.

KHUS
[*Vetiveria zizaniodes* (Linn) Nat]

S.D. UPADHYAYA
Professor
Department of Plant Physiology, J.N.K.V.V. , Jabalpur

Botanical Name	:	*Vetiveria zizaniodes*
Family	:	Poaceae
English Name	:	Vetiver
Indian Name	:	Sugandhimulah (Sanskrit), Khas-Khas (Hindi).
Distribution	:	India, Tropical Asia, Africa, Australia, Brazil, Congo, Haiti, Indonesia, Jamaica etc.

Uses

Soft drinks, chewing tobacco, paan Masala, Soaps, perfumes, Cosmetics, cooler.

Introduction

Khus is a densely tufted, perennial grass, found growing on various types of soils. The under ground part of the plant consists of numerous fine rootlets of light-yellow or gray to reddish in colour, which contain a viscous essential oil, with a pleasant and persistent odour. The commercial oil of vetiver is obtained by the distillation of the root. Due to the viscous nature, the vetiver oil serves as a fixative and imparts a pleasing and lasting note to the perfume composition.

The total world production of vetiver oil is estimated about 100 tons annually (Haiti, Indonesia, China, India and Brazil). In India, about 20-25 tons of oil is produced annually and yet its demand for perfumes, essences, attars and soaps etc. is greater than indigenous production. In quality Indian vetiver oil, especially of North Indian origin is considered to be best in the world market.

High demand of the vetiver oil extracted from the plants growing in the wild has been fast depleting the wild resources of vetiver. The selling rates of roots and oil are high. The cultivation of genetically improved vetiver on large scale can meet the needs of industry for vetiver roots of reasonable price.

Distribution

Vetiver is indigenous to the Indian subcontinent and is widely distributed in tropical Asia, Africa and Australia. It flourishes on the slopes of the Himalayan Mountains and in many parts of India, Sri Lanka and Malaysia. In India, it is found growing wild in the forest of Rajasthan and Uttar Pradesh. It is also found in some parts of Assam, Jammu, Bihar, Orissa, Andhra Pradesh and Karnataka. However, it is systematically cultivated as a crop in Kerala. Its cultivation is picking up in parts of Madhya Pradesh, Uttar Pradesh and Haryana.

Soil and climate

The vetiver grows luxuriantly in the places with an annual rainfall of about 1000-2000 mm, temperature ranging from 70° F to 110° F and with a moderately humid climate. Semi-temperate and temperate climate is not fit for its cultivation.

Vetiver plant grows in any soil but loamy cum sandy-soil is most suitable. Warm and damp climate and frequent irrigation or rains are the two main requirements for its successful growth and cultivation.

Vetiver plant can be grown on problem soils, including saline and sodic, sandy, water logged, high pH (pH 9.5 to 10.0), sloppy and undulated soils (for soil conservation purpose).

Preparation of land

Vetiver field should be free of perennial weeds and shrubs, which can be obtained by two-three deep ploughing, preferably during summer months. During planting season the land is ploughed, harrowed and brought to a fine tilth. The land is laid in to plot beds of convenient size.

Description of the plant

It is a perennial grass. The clumps are erect, growing to a height of 1 to 1.8 m. The underground stem is branching with a spongy, aromatic root system having fine rootlets, 10-11 cm long and 1-3 cm. in diameter. The aerial stem is tall, smooth, glabrous and the leaf is polished. The leaves are long with the leaf-blade 30-40 cm. long erect rigid, firm, narrow, 0.7 cm, or less wide and somewhat spongy; usually glabrous, pale-green and rough along the edges. The panicle is an oblong of many rayed whorls of slender simple racemes (and is terminal), 5 cm long with pairs of dissimilar spike lets, the pedicel is male and the sessile is hermaphrodite, the spike lets are reddish brown or purplish. The anthers are 2 mm in size and the stigma is purple in colour. The grain is oblong and oblique at the top. The pedicelled spikelets, which are male, are smooth; the upper floral glume is etitire, acute and 3 mm long. Vetiver is an autogamous plant. Self-pollination does not occur and it is an obligatory cross-pollinated crop.

Improved varieties

The grass growing wild in North India (mainly the flowering type) and that cultivated in South India (non-flowering type) are two different types. North Indian type is poor in oil yield, but its oil quality is superior; whereas the South Indian type yields more oil, but the quality is inferior. A number of varieties of vetiver are available for commercial cultivation.

CIMAP has developed two superior clones KS-1 and KS-2 (collection of North Indian type). These have a yield, which is almost comparable to the existing high-yielding material. However, the quality of oil from both the clones was rated the best among all the clones evaluated.

Improved tretraploid strain named 'Sugandha' has also been released by the CIMAP and is superior in terms of oil-yield over the corresponding diploid parent. Strain yield up to 1.4% oil content in fresh roots and 21.2 q/ha of roots, against 1.15% oil and 18.5 q/ha of roots in the control.

Pusa hybrid numbers (1-30) have been evolved by the NBPGR, New Delhi. F. hybrid, Hyb-26 gives higher yield of roots (14. 58 q/ha) with an oil content of 1.50%. Hybrid 26, 7 and 16 perform better in saline and alkaline soils with high root and oil-yield.

- Under Bang lore conditions, hybrid-8 has been found to be better. Its roots yield 1% essential oil and possess 70-80% vetiverol content.
- A vetiver clone, ODV-3, developed at the lemongrass research station, Odakkali, is reported to give good root and oil-yield in Kerala.

Cultivation

Propagation

Vetiver is propagated from seeds or slips. Seeding is found to be profuse in the crop in areas where it occurs in the wild and regeneration takes place from self-sown seeds and by tilling during and after the rains. In South-India, where the crop is cultivated on a large scale, the non-seeding type is grown. This is propagated form slip obtained from the uprooted clumps of the previous crop.

Planting

Slips of vetiver can be planted any time between February and October, but July and August is considered most ideal time. Rainy season is also ideal for un-irrigated areas, slopy/undulated and problem soils. Vetiver plants are cut 25-30 cm above ground level and clumps are dugged and splitted into slips for out planting. Vetiver is planted in rows 45 cm apart with a plant to plant distance of 30 cm. A spacing of 60 x 25 cm is also recommended to allow 60,000 plants/ha to produce maximum roots.

Manures and fertilizers

Timely manuring is essential for increasing yield of roots and oil. In vetiver crop field 10 t FYM, 25 kg/ha each of N, P and K are applied. While the entire quantity of FYM, half N and a full dose of P and K are applied at planting, the remaining half of N is given after about 6 months of planting. A Fertilizer dose of 60 kg N, 22.5 each of P_2O_5 and K_2O is recommended in kerala.

Irrigation

Frequent irrigation are essential for February, March, April planted field and till the monsoon starts. In monsoon season one light irrigation is necessary after planting if there is no rains.

Weed control

The initial growth of vetiver for first 60-70 days is very slow and this provides infestation of a number of weeds, which compete with it for nutrients and moisture and reduce oil yield about 75% over a weed free field. This weed growth can be checked by use of herbicides (atrazine @ 0.500 kg a.i./ha for Kharif weeds) or intercultural operations (three-four weeding are essential during first and one or two in the second year).

Pests and diseases

Very few diseases have been reported on vetiver. During the rainy season, the plant is attacked by fusarium spp. It can be effectively controlled by drenching the soil with Bordeaux mixture. Leaf-blight, caused by Curvularia trifolii can be controlled by repeated spraying of copper fungicides in appropriate doses.

There are no serious insect pests, which attack this crop. Some times, white grubs and termites damage the crop by cutting the roots. These can be controlled by broad casting 65% lindane (125 kg/ha) before the last ploughing.

Removal of vegetative parts/ flowering inflorescence The vegetative and flowering shoots are removed 20-30 cm above the ground for better tilling, root growth and better accumulation of oil in roots as following manners and time: -

(i) First removal of vegetative portions: After 4-5 months of planting.

(ii) Second harvesting of areal shoots: Before flowering of plants.

(iii) Third harvesting of areal shoots: One month before digging of roots during winter season.

Digging of roots (harvesting)

The time of harvesting is very important, as the yield of roots and oil percentage vary with changes in environmental conditions. Depending upon time of planting, roots are dug at 15 months age. Digging is always done during winter months (Nov-Feb.), when the oil concentration in roots reached to the maximum and quality of oil is best. Digging of roots after February month may lead to low oil recovery and poor quality of oil.

Harvesting is usually done manually by digging out the bush along with its roots. On large-scale cultivation, for harvesting the vetiver roots, a single disc of disc plough mounted on a tractor was used, which uprooted the roots from 30-35-cm. depths and turned these in field. This method is useful for saving manpower and also able to give 15% higher root recovery over manual digging. The roots are then separated from the leaves, washed and dried under the shade for 1-2 days before distillation, This improves the quality of the essential oil while prolonged drying in the sun reduces the yield.

Yield

Depending upon climate, soil and management practices, root yield vary from 10-50 ql/ha. The average yield is about 25-30 ql/ha. In commercial field oil recovery on commercial scale ranged from 0.5-0.8% and 20-30 kg/ha. Average yield of vetiver roots and oil in various soils are.

S.No.	Type of soil	Root yield (ql/ha)	Oil yield (kg/ha)
1.	Fertile (loamy)	45-50	25-30
2.	Medium Fertile (Sandy-loamy)	35-40	20-25
3.	Sandy	10-15	8-10
4.	Water logged	15	10-12
5.	Alkaline	10-15	8-10

Oil distillation

The oil is extracted by steam distillation from both dry and fresh roots. The oil obtained from stored roots is more viscous and possesses a slightly better aroma than that obtained from freshly harvested roots.

The roots should be distilled when fresh, in order to obtain the maximum yield of oil and to reduce the time of distillation. The roots are cleaned and steeped in water for 12-16 hrs. Chopped pieces 2-4" long are charged into the copper/stainless steel still for 12-15 hrs. The still operate on the principle of water and/or steam distillation. The distillation of roots is generally done in country stills. In North India, the distillation process is completed in 12-14 hours while in South India, it usually takes 72-96 hours.

The oil yield of vetiver roots varies on a number of factors such as: the degree of dryness of the roots, the age of the roots, the nature of the land under cultivation, the lifting time, the perfection of the distilling equipment and so on.

Ageing for a period of six months improves the odour of the oil substantially. The 'harsh' 'green' and 'earthy odour' character of the freshly distilled oil disappears and a fuller, heavier and sweeter odour develops.

ROSES

A.C. VERMA
Professor and Head
Deptt. of Horticulture, College of Agriculture, JNKVV, Jabalpur

Botanical Name : *Rosa indica*
Family : Rosaceae
Common Name : Gulab

Roses

Roses are symbol of beauty, fragrance, and are used to convey the message of love. Without roses a garden is not considered as a complete garden. A great diversity in plant growth, colour of flowers, flower shape, fragrance, slow opening of flowers and good keeping quality has made this flower so popular that it is grown as a commercial flower. It is a top ranking cut flower in the flower trade in India.

In India roses are grown for cut flowers. It is also used for making garland, guldasta (bouquets) Jhalar, Chadar, wreath, buttonhole. Flowers are also used for making essential oil, rose water and gulkand. Dry petals of roses are used for making incense sticks. Besides, it is also used for making medicines for heart diseases and heart tonic. Powder of rose fruits is used for treating the wounds. Extract of rose is also used for killing bacteria of tuberculosis and Diarrhoea in human beings.

Varieties

There are about 150 species of roses but very few are popular and cultivated in India. Based on their flowering behaviour, size of flower and growth habit the roses are classified into four different groups:

Classification of roses

1. Hybrid Teas roses (H.T.)
2. Floribundas roses
3. Climbers roses
4. Miniature roses
5. Shrub roses

Hybrid Teas (hybrid perpetuals × Tea roses):

Characteristics

- It produces flowers on long canes.
- It produces single elongated buds with slow opening character
- Buds have good keeping quality

Commercial varieties

1. Bedding varieties

Yellow: McGredys' Sunset, Golden Giant, Dutch Gold

Orange and Scarlet: Super Star, Fragrant Cloud and Montezuma.

Orange Yellow : Raja Ram Mohan Roy and Bettina

Pink : Confidence, South Seas, Firt Prize and Sonia.

Red Dark Crimson : Christian Dior, Avon, Crimson Glory, Oklahama,

Mauveor Lavender: Blue Moon, Sterling Silver and Paradise

White: June Bride and Tusar

Bi- Colour: Bajazzo, Oriana, and Flamming sunset

Colour blend: Careless love, Kiss of Fire and Double Delight

Stripped colour: Anvil Sparks, Careless Love and Siddhartha

2. *Exhibition varieties:* Blooms of these varieties are large and have good vaselife

Anvil Sparks, Avon, Christian Dior, Effel Tower, First Prize, Garden Party, Montezuma Pusa Sonia, Show Girl and Super Star.

3. Scented Varieties

Avon, Blue moon, Crimson Glory, Sugandha and Double Delight.

4. Commercial varieties for cut flowers

Super Star, Montezuma, Pusa Sonia ,Happiness, Double Delight and Alec's Red.

Floribundas

- These Varieties produces flowers in clusters smaller than the Hybrrid Teas.
- Flowers are produced in great profusen apearing like bouquets.

Important varieties

Red : Marlena, Gabriella and Jantar Mantar

Orange Scarlet: Independence day, Shola and Zambra

Yellow All Gold, Sunflare, Gold Buny.

Pink: Queen Elizabeth and Junior Miss

White: Iceberg and Himangani.

Lavender: Angel face and Lavender Princes

Multicolour: Charleston, Banjaran and Madhura.

Climbers Roses

These roses are most suitable for arches and pillers and produces magnificent effect.

Important varieties

Fragrant Cloud, Golden Shower, Royal Gold, White Pearl, Angel Face and Paradise.

Miniature Roses

These Miniature roses were introduced from China under the name as Pigmy roses. In India , Miniature roses are also called as baby or fairy roses.

Characteristics

- It has long blooming period
- It is least affected by cold and diseases
- These are most ideally suited for edging, pots and rockeries
- These are easily propagated by semi hard and hard wood cuttings

Important varieties

Red: Redflush and Desert charm

Pink: Cupcake and cuddless

Orange: Sunblaze, Marry Marshal and Chattem Centennial

Yellow: Yellow doll, and Summer butter

White: Cindrella, Green Ice and Misty Dawn,

Multicolour: Party Girl and Puppy love

Bicolour: Dream Glow and Glory Glow.

Climate

It requires moderately cool climate with bright sunshine and open place. Plant growing under shade produces less flower yield. Most of the cultivars grow best at temperature ranging between 15-25^{0}C With good quality and quantity of flowers.

Soil

The ideal soil for roses is medium to loam having sufficient organic matter. Soil should be well drained and have a pH value of 6.0-7.0.

Methods of propagation

There are two methods

1. By shield budding 2. By cutting

Hybrid Teas (HT), Floridundas and Climbers Group of roses are commonly propagated by 'T' or shield budding.

Miniature roses and Deshi roses (Khooni Gulab) are generally propagated by semi hard and hard wood cutting.

Budding height

Based on the height on which budding on the root stick is done Budded roses are grouped into three groups

- *Bush roses:* budding is done at the 7.5- 10 cm above the ground level
- *Standard roses:* Budding is done at the height of 100 cm from the ground level.

  These standards are more useful in both formal and informal types of garden.
- *Half Standard:* These are prepared by budding at a height of 50 cm from the ground level.
- *Weeping Standard:* weeping Standard roses are prepared by budding of climbing roses at 100 cm above the ground level. In weeping roses branches of climbing rose are droop down and create artistic effect.

Time of budding

The best time for budding roses is November to February. There is a 90-95% success of bud if 'T' or shield budding is practiced during the month of November to February.

Suitable root stock

The most commonly rootstock used in propagation of roses are :

1. Edourd (*Rosa bourboniana* or *R. multiflora*)
2. *Rosa indica* var.odorata

Preparation of land

- The plot should be dug thoroughly to a depth of 90-100 cm and it should be kept open for few days.
- The soil should be dried and refilled with 10-15 kg well rotten FYM per square meter.

Planting time: End of September to mid of October or it can be extended upto mid of November.

Spacing

HT roses - 75 × 75 cm

Floribundas and other groups - 60 ×60 cm.

Manures and fertilizers

Manure/ Fertilizer	Dose	Time and Method of application
Well rotten FYM	8-10 kg/plant or 30 t/ha	FYM should be applied around th e plant in basin just after the pruning
NPK	50-100g N/m^2 100-200g P/m^2 50-75g K /m^2	These fertilizer should be applied in two split doses. 1/2 quantity of N full dose of P and K at the time of pruning and remaining half dose of N should be applied 1 month after the first application. Potash should be given in the form of Potassium sulphate. It should never be applied in the form of muriate of potash as gives adverse effect on the quality and colour of flower.

Foliar feeding

For better effect and good quality of flower, foliar feeding of roses has been found very useful. Foliar spray of a mixture of nutrients containing **urea, dihydrogen ammonium phosphate, potassium nitrate and potassium phosphate** is used commercially to improve yield and quality of flower.

Ratio of nutrient mixture for preparation of spray solution :

Nutrient	Quantity (g)	Spraying
Urea	12	Foliar spray of 30 g mixture /10 litre of water is given at weekly or fortnightly interval for good quality of flowers.
Dihydrogen ammonium phosphate	6	
potassium nitrate	6	
potassium phosphate	6	

Foliar feeding with micronutrients

Foliar feeding with micronutrient in roses is also done for improving quality of flowers. The commercial formulations of these micronutrients like tracel-2 and Multiplex etc. are available in the market, which can be used safely.

Irrigation

Water requirement of roses depends upon soil types and season. Light soils required more frequent irrigation than the heavy soil. During winter season irrigation is done at about 7-10 days interval whereas during summer it should be done at a interval of 5-6 days.

Pruning

It is most important and technical aspect of rose growing if pruning is not done properly plant produce poor quality of flower and becomes unproductive.

Time of pruning

It depends entirely on the climatic conditions of the region

Temperate climate: March-April

Tropical and Sub-Tropical climate: Middle of October - First week of November

Method of pruning : Always make a sharp cut at an angle of 45^0 about 5mm above a strong healthy eye pointing outward.

Type of pruning

Heavy pruning: In HT roses heavy pruning is practiced leaving 4-6 strong canes on all sides of the bush cutting back about 30-45 cm above the bud union.

Light pruning: Floribundas group of roses are given light pruning.

Thinning: Miniatures are given very light pruning by doing thinning of over crowding branches.

Spray of fungicides and insecticides should be give immediately after the pruning.

Harvesting stage

The flowers for decoration and dispatch should be cut at tight bud stage when buds show their colour, but the petals have not yet started unfurling. For making garland of roses (Khooni gulab) flowers should be cut at full bloom stage.

Important diseases and pest of roses

Disease	Symptoms	Control measures
Die back (*Diplodia roseum*) (*Colletotrichum spp*)	Blackening of the twigs starts from the cut ends of the pruned plants and proceed downwards. The severity of these diseases is noticed after the pruning operation after the month of October.	1. Removal of affected parts about an inch below in the healthy part. 2. Spray of carbendazim @ 1 g/l of water. Or drenching of soil with solution of carbendazim.
Powdery mildew (Sphaerotheca *pannosa* var.rosae)	The disease appears during February- March when temperature starts rising. Whitish powder on upper surface of the tender leaves and tender growth appears. Affected flower buds do not open and show discolouration of petals.	Fungicides like sulphur dust, bavistin and calaxin have been found most effective in controlling the disease.
Black spot (*Diplocarpon rosae*)	Dark brown circular spots are seen on the leaves. These spots multiply, expand and join together and entire leaf turn yellow and latter fall down. It becomes more severe during rainy season.	Spray of Zineb @ 0.25% at regular interval of 15 days can reduce the incidence of disease.
Insects		
Red scale	It is a very serious pest of rose, which attacks mostly in August and September. The branches are covered with reddish brown incrustations under which the insect sucks the juice of the plants. The plants may dry up completely under severe attack.	Spray of malathion or rogor @ 0.25% may reduce the incidence of insect.

MARIGOLD

A.C. VERMA
Professor and Head
Deptt. of Horticulture, College of Agriculture, JNKVV, Jabalpur

Botanical Name : *Tagetes erecta*
Family : Asteraceae
Common Name : Genda

Marigold is one of the most commercialy grown flowers in India. It has gained popularity amongst gardeners and flower dealers on account of its easy culture and wide acceptability. In India, it is grown in about 13000 hectares. Commercial growing of marigold is done in Uttar Pradesh, Bihar, Haryana, Punjab, Rajasthan, Madhya Pradesh, West Bengal, Maharashtra, Karnataka, Tamil Nadu and Andhra Pradesh.

Uses

1. Fresh flowers of African and French marigolds are used for religious offerings and social functions.
2. Marigolds are grown for the extraction of carotene pigments like Xanthophylls from the flowers petal and is used for poultry feed..
3. Marigold oil has a broncho-dilatory, tranqualizing and anti-inflammatory property.
4. The oil of marigold is also used as insect repellent activities against flies, ants and mosquitoes.
5. Oil of the marigold is also used for preparation of high grade perfumes and cosmetics.
6. Floral extract is used as a remedy for various eye diseases and ulcers.
7. Extract of petal is also used with equal quantity of Ghee in treating bleeding piles in human being.

Varieties

The genus Tagetes includes 33 species in which Tagetes erecta (African marigold) and T. patula (French marigold) are of great horticultural importance and are grown commercially for their flowers.

1. African marigold

Pusa Narangi Gainda, Pusa Basanti Gainda, Alaska, Apricot, Double Eagle.

F-1 hybrids : Climax, Apollo and Moon shot.

2. French marigold

Single: Dainty Marietta, Legion of Honour and Sunny.

Double: Bolero, Bonita, Carmen, Fiesta, Orange Flame Red and Gold butterscotch.

Climate

Marigold requires mild climate for luxuriant growth and profuse flowering. It ceases to grow at high temperature thereby flowers quantity and quality are adversely affected. During severe winter including frost plants and flowers are killed and blackened under North Indian conditions. It needs plenty of sunshine and are grown in open sunny situation.

Soil

Marigold can be successfully cultivated on a wide variety of soils. However, a well-drained, fertile and deep soil having good water holding capacity is well suited for marigold cultivation. The soil should be near to neutral in reaction viz., pH 7.0-7.5.

Soil preparation

Land should be well prepared by ploughing it 2-3 times and 100 q/ha of well rotten FYM should be well mixed. Beds of convenient size be made to facilitate irrigation and other cultural operations.

Raising of seedling

Raised nursery beds of 1mx 3m should be prepared with adding well rotten FYM. If soil is heavy sand may be mixed to make the soil of good drainage. To minimize the mortality of plant nursery beds should be drenched with 0.2% Captan before sowing. Seed sowing should be done when the temperature ranging from 18 to 30^{0}C. As soon as the beds are ready seed should be shown in lines spaced at 6 to 8 cm. Seeds may be sown at a depth of 2 cm., and then covered with a thin layer of well rotten FYM. Seed rate per hectare is 800 g to 1 kg.

Planting

The seedlings are transplanted in the field after one month of seed sowing at a spacing of 40 x 40 cm for African marigold and 30 x 30 cm for French marigold. During summer and rains transplanting should be done in the afternoon or during cloudy weather. The terminal portion of plants should be pinched as and when plants establish well in the field. This well help in the production of more side shoots.

Sowing and transplanting time:

Season	Sowing time	Transplanting time
Rainy	End of June to 1 st week of July	First fortnight of August
Winter	Mid of September	**Mid of October**

Manures and Fertilizers

Application of 100 kg N, 100 kg P_2O_5 and 100 kg K_2O along with 10 t FYM is advised for production of good quality flowers of Marigold. Whole quantity of phosphorus and potash as well as FYM should be applied at the time of land preparation whereas the nitrogen should be applied in two split doses after one month of transplanting.

Foliar feeding

It will be better if two foliar sprays of 0.2% urea are done at an interval of 15 days.

Irrigation

Rainy season crop is irrigated as per the atmospheric conditions. Winter crop needs watering at 8-10 days intervals depending upon the water requirements of the plants.

Harvesting and packing

The flowers are harvested when they are fully open. After harvesting, the flowers are packed in small and big bamboo baskets. They are then transplanted to nearest or distant markets by buses or lorries. It is better to harvest the flowers in the morning and to store them in a cool place before packing.

Pinching of marigold plants

In tall varieties of *Tagetes erecta* emergence of side branches and their flowering is influenced by the presence of apical dominance. If the terminal portion of shoot is removed early, emergence of side branches starts earlier and more number of flowers of good quality and uniform size are produced. Results of studies on pinching revealed that pinching the plants 40 days after transplanting enabled the plants to yield more flowers. However, the plants remained dwarf and flowering was delayed in comparison to late pinching i.e. 50 or 60 days after transplanting.

Flower yield

If proper cultural practices are followed, an yield of 20-22 tonnes of fresh flowers can be obtained from one hectare plantation of African marigold, while from French marigold it is 10-12 tonnes per hectare. Yield of essential oil is 50-60 kg/ha.

Insect- pests and diseases

Insect	Nature of damaged/symptoms	Control measures
Red spider mites	The mites make web on the foliage and suck the sap from tender portions.	Spraying the plant with 0.2% malathion or monocrotophos @ 1ml/lit of water.
Hairy Caterpillar	The caterpillar eats way foliage and tender parts of plants.	The caterpillar can be controlled by sprays of Endosulphan @ 2 ml/Lit water.
Diseases		
Powdery mildew	The affected plants pa rts shows whitish powder on leaves and stems.	Spray 0.2% wettable sulphur. or Karathane 40 E.C. @ 0.5% at fortnightly intervals.
Leaf spots	Minute brown, circular spots appear on lower leaves and enlarge at later stage leading to premature defoliation and ultimate death of the plants.	Crop should be sprayed with Mancozeb @ 0.2% at fortnightly intervals.

TUBEROSE (*Polianthes tuberosa* L)

A.C. VERMA
Professor and Head
Deptt. of Horticulture, College of Agriculture, JNKVV, Jabalpur

Botanical Name : *Poliantes tuberosa*
Family : Agavaceae
Common Name : Gulshava

It is most important commercial flower grown in the states of Karnataka, Tamil Nadu, Andhra Pradesh, Maharashtra, West Bengal, Orissa, Gujrat, Uttar Pradesh, Haryana and Delhi. It bears waxy, white and fragrant flowers in pairs on long spikes. It is mainly grown for use as cut flowers, loose flowers, pot flowers and extraction of essential oils. Apart from domestic consumption it has very good export potential also.

Uses

i. Grown for cut flowers.
ii. For loose flowers.
iii. For pot flowers
iv. For extraction of essential oil.

Cultivars

Four types of tuberose are grown in India.

1. **Single type:** Its florets have only one row of petals which are pure white in colour. It gives highest quantity of essential oil recovery. It is commercially grown for use as cut flowers, loose flowers and extraction of essential oil.
2. **Semi double type:** The florets have 2-3 rows of petals. It is grown for cut flower purpose.

3. **Double type:** The florets have more than three rows of petals. Which are white in colour, but tinged with pinkish red. It is mainly cultivated for use in cut flower.
4. **Variegated type:** Its florets are similar to cultivar single except that their tinged have pinkish colour.

Improved varieties

i. Shri Nagar (single type) Rajat Rekha (single type)
ii. Svarna Rekha (Double type)
iii. Suvansini (Double type)

Climate

It is a tropical flower and grows well under moderate climatic condition. It is also well adopted to sub tropical conditions. The optimum temperature for its proper growth and development is 20-30^{0}C. It prefers open sunny location and gives poor yield in shady places.

Soil

Loam and sandy soils having a pH of 6.5 to 7.5 with good aeration and drainage are ideal for its successful cultivation.

Planting of bulbs

Tuberose is mainly propagated by means of bulbs and bulblets. For obtaining regular flowers for longer duration, planting of bulbs at 10-15 days intervals may be done.

1. **Planting time** – April to June
2. **Size of seed bulbs** - 2.0 cm or above.
3. **Depth of planting**- 4.0 to 8.0 cm
4. **Spacing** - 20-10 cm plant to plant, 30-20 cm row to row.

Irrigation

It is advisable to irrigate the field before planting of bulbs and thereafter further irrigation could be avoided until the bulbs have been sprouted. Usually during summer months the crop should be irrigated at 5-7 days intervals and during winter months at 10-12 days intervals.

Application of manures and fertilizers

FYM : 250 –300 q/ha

Fertilizers : 100-250 kg N/ha
100-150 kg P_2O_5 /ha
75-100kg K_2O/ha

Methods of fertilizer application

1. Full dose of phosphorus and potassium should be applied as basal dose at final preparation of plots.
2. Nitrogen should be applied in three split doses at planting time, 30 and 60 days after planting.

Use of bio-fertilizer

Bulbs inoculated by dipping for a minutes in thick slurry solution of biofertilizers viz Azotobactor, Azospirillum or mixture of both gave 19-21% increased yield of flower spikes and also significantly increases bulb and bulblets production.

Harvesting of flowers

Tuberose comes in flowering after 3-5 months of planting. The whole spike is cut from the plant for cut flower purposes whereas for loose flower and essential oil extraction only individual florets at full maturity stage are picked up. The spikes are cut with a sharp secateur leaving 4-5 cm pace on the plant, so as not to damage the growing bulbs and immediately place them in water.

Packaging and post harvest care of flowers

The loose flowers are packed in bamboo baskets, holding 15-20 kg florets each and are transported to the nearby market. The flower spikes are graded according to their total length, length of rachis and quality of individual florets and then bunched in round bundles, each having

50-100 spikes. The stem portion of bundles is wrapped in wet newspaper and later the bundles are wrapped in soft tissue paper or polythene film to avoid any damage to the flowers. These bundles may be packed in cardboard boxes and transported to the destination as quickly as possible. The optimum storage temperature for cultivars Singe and Double spikes is 7^0 C at which spikes can be stored for five days. Pulsing of spikes at low temperature (10^0C) for nearly four hours with 4 per cent glucose solution is helpful in prolonging their vase life.

Lifting of bulb

i. Bulbs attain maturity when flowering is over and plant growth ceases.
ii. Drying of old leaves is also an indication of bulbs maturity.
iii. Irrigation is withheld a few days prior to harvesting of bulb.
iv. Bulbs are dugout with a spade without soil.

Yield

i. Normally 2-4 lakhs spikes /ha are harvested.
ii. 10-15 tones loose flowers / ha.
iii. 20-30 tones bulb/ha will give additional income.

Storage

i. After harvesting bulbs are graded to different grades depending upon their size.
ii. Different grade of bulb are stored separately in a cool, dry and shady place.
iii. During storage, bulb should be stirred every 8-10 days to check the spread of mould and rot.

AGROTECHNIQUE AND MEDICINAL IMPORTANCE OF SOME COMMON SPICES

SUBRATA SHARMA

Deptt. of Plant Breeding and Genetics, JNKVV, Jabalpur.

The fame of Indian spices is perhaps older than our recorded history. Since antiquity, India has been a leading spice producing and exporting country in the world. The world still looks upon India as the 'home of spices'. Spices constitute an important group of agricultural commodities, which are indispensable, as they are used in flavoring foods, pharmaceutical, perfumery, cosmetics and several other industries. Some spices like clove and mustard are also used for preserving food. They also form a means of earning foreign exchange for our government. Besides, huge quantities of spices are consumed within our country for seasoning of foods and various other purposes.

The agrotechniques of some important spices cultivated in this region are discussed here.

Bishops weed or Ajwain *(Trachyspermum ammi)*

The plant produces attractive umbels and the seeds have medicinal value. The seeds contain active principle '**Thymol**' and '**Visnagin**'. It is selective coronary vasodilator and bronchidilator and also used to treat coronary insufficiency and bronchial asthma. Decoction of fruits is used to treat urethral spasm, to cause passage of kidney stones and renal colic. India earns foreign currency by exporting Thymol.

Varieties

NPD-30, NPD-79, NP(J)-8, NP(J)-66.

Climate

It is a Rabi cultivated crop and can be successfully cultivated under temperate and subtropical conditions. It requires cool climate for vegetative growth and warmth with long day length during flowering and ripening.

Soil

Sandy loam to loam soil is best suitable for cultivation of ajwain. Acidic or basic soil is not suitable for cultivation of this crop. It gives good yield in soils having pH range of 6.5 to 8.2.

Inputs

Seeds	-	3.0 to 4.0 Kg / ha.
FYM	-	20 to 25 tons / ha.
Super phosphate	-	250Kg / ha.
Muriate of potash	-	60 Kg / ha.

Cultivation

The land should be well tilled, harrowed three to four times and leveled before sowing seeds. The seeds are either broadcasted or sown in lines. To get good yield, sowing in lines of 30x20cm. is recommended. The seeds are small and are mixed with fine soil or sand in ratio of 1: 10 to maintain proper plant population. The plants are further thinned at four to six leaf stage.

Irrigation

The crop is irrigated initially 2-3 days after sowing. Flood irrigation should be avoided for proper germination of seeds. Later, 6-8 more irrigation are required. When seeds start ripening, irrigation should be stopped.

Interculture

Three to four hoeings are done to keep the field weed free. Weedicides like 2-4D @ 2.4Kg dissolved in 700 litres of water and sprayed / ha is effective for controlling weeds.

Harvest

The crop is harvested 195 to 240 days after sowing when the umbels have turned brown. The umbels are harvested by hand. The product is shade dried and seeds separated manually by beating with sticks and winnowing.

Yield

7-8Q seeds / ha.

Plant protection

Major insects

White ants and cut worms.

Control

Spray with 0.1% solution of Metasystox.

Major diseases

Damping off, powdery mildew and wilt.

Control

Spray Mencozeb difolton solution @ 0.2% at 30-35 and 40-45 days after sowing. If needed 25 Kg sulphur can also be added.

Cumin (*Cuminum cyminum L.*)

Cumin is an annual herb and cultivated throughout India for aromatic fruit. The seed yields volatile oil of 2.4 to 3.6%. The principle content of this oil is cuminaldehyde which ranges from 35 to 65%. The cumin oil is used as a flavoring agent in curries and culinary preparations. It is also used in medicine. The fruit is pungent, hot, sweet, cooling, carminative, antidysentric, stomachic and uterine stimulant. It is used in medicine to cure eye disease, increase appetite, cure leucoderma, leprosy, asthma, gonorrhea. It also relieves hiccough, inflammation, applied to balls and ulcers (Unani) medicine. It is also used in preparing medicine for snakebite and scorpion sting.

Varieties

Pusa 33, 66 and 90, SC-43.

Inputs

Seed rate: 8-10Kg / ha

Soil, fertilizer dose and cultivation technique

Same as bishops weed or ajwain.

Yield

8-10Q/ha.

Turmeric (Curcuma longa Linn.)

Turmeric is one of the most important and ancient spices of India and is used daily extensively by all classes of people as it adds a typical flavor and colour to the dishes. Besides, it is also use as a dye in cotton textiles, in medicine and cosmetics. Its principle ingredient is 'Curcumin'which is in great demand abroad and a source of foreign exchange for our country. It is extensively used as a food colorant and dye all over the world. The turmeric oil has antiseptic properties. The juice of raw rhizomes is used to cure skin ailments and leprosy. It is a stomachic, carminative, tonic and blood purifier. Apart from this it is consumed in cosmetic preparations and considered sacred and used in religious and ceremonial functions by the Hindus.

Varieties

Analapuram, Soni, Gautam, Rashmi, BSR-1 and Suverna.

Climate

It is warmth requiring crop and cultivated at almost sea level to 500 meters above sea level. Temperature below 220 C is unfavourable for growth of the crop. Low initial rainfall during sowing, moderate rainfall during growth period and dry climate one month prior to harvest of the crop is best suitable for cultivation.

Soil

It grows on light black, ashy loam and red soil to stiff black clayey loams in irrigated and rainfed areas. However it thrives best in a well drained sandy soil or clayey loam soil. The crop cannot withstand salinity or alkalinity.

Inputs

Rhizomes

Healthy, well developed, well dried and disease free rhizomes and fingers are used as planting material. Each finger should have 2-3 eye buds. Prior to sowing, the rhizomes should be treated with 2% solution of Agallol.

Seed rate

12-15 Quintal rhizomes / ha.

FYM

300 Q / ha Ammonium sulphate - 200Kg / ha

Muriate of Potash - 200 Kg / ha

150 Kg more Ammonium sulphate should be added 40 days after sowing and 65 Kg urea as top dressing 60-80 days after sowing.

Cultivation

The land is prepared with the receipt of early rains. The soil is brought to a fine tilth by giving 4-6 ploughings. For planting, beds of about 1-1.2metres width, 1.5cm height and convenient length are prepared, leaving spacing of 40 cm. between the beds. Planting is done by forming ridges and furrows. The rhizome are placed in the furrows and covered with soil, maintaining distance of 20-25cm. between rows and 15-20cm. between plants.

Mulching

The crop may be mulched immediately after planting with green leaves @ 12500 – 15000 kilograms / hectare. It may be repeated after two months with same quantity of green leaves.

Irrigation

15-20 irrigations are given depending upon the soil.

Interculture

3-4 hoeings are necessary depending upon the nature of the soil and weeds.

Harvest

The crop becomes ready for harvest 9-10 months after sowing when the leaves senesce and turn yellow. The land is ploughed and rhizomes are gathered by hand picking or the clumps are carefully lifted with a spade.

Yield

200-250Quintals / ha raw turmeric.

Curing

The fingers are separated from the harvested rhizomes and form the seed material for the next planting season. The rest of the rhizomes are then cleaned and boiled in water in copper, galvanized iron or earthen vessels. Stop boiling when froth starts coming out and white fumes come out. When cooked properly, the rhizomes feel soft when pressed between the fingers. The boiled rhizomes are then spread in the sun for 10 to 15 days till they become hard and brittle. They are then cleaned and polished mechanically in a drum rotated by hand or in a wooden drum rotated by power.

Colouring

Boiled, dried and half-polished turmeric is taken in a bucket, which is shaken continuously with a prepared emulsion poured in. When the fingers are coated with emulsion they may be dried in the sun.

Emulsion

Alum	0.04kg.
Turmeric powder	2.00kg.

Castor seed	0.14kg.
Sodium bisulphite	30gm.
Concentrated	
Hydrochloric acid	30 ml.

Yield

4-5 tons cured turmeric / ha.

Plant protection

Pest

Shoot borer, leaf roller

Control

0.05% Malathion or Dimethoate at time of planting or when required.

Disease

Leaf spot, rhizome rot.

Control

Treat seed rhizomes before planting with Agallol (0.5% for 3 minutes), wettable cerason (0.1% for 30 minutes). Application of chestnut compound is recommended when disease occurs in the field. It is prepared by mixing well powdered Ammonium carbonate and Copper sulphate in ratio 11:2.

Ginger: (*Zingiber officinale*)

Ginger is grown throughout India and contributes about 50% of the total world requirement. India is the biggest producer and exporter of ginger in the world. Kerala alone contributes about 40-50% of the total production of ginger in India.The rhizomes are sold raw or after curing as 'sonth'. It is used in food as Murabba, candy, pickles and for preparing ginger cocktail and ginger wine to aid digestion. Ginger oil (Oleoresin) contains gingerol, Zingerone, resins, phenols and volatile oil, which is, used as a flavourant in soft drinks, as a spice in bakery products and in pharmaceutical use as carminative, stimulant in gastritis, dyspepsia, flatulent colic etc. In ayurveda it is used to relieve cough and vatha.

Varieties

Surabhi, Suruchi, Suprabha.

Climate

Same as *Curcuma longa* Linn.

Inputs

Seed rate

20-25Quintals of rhizomes / ha.

The rhizome pieces having 2 to 3 eye buds each are treated with 0.25% Dithane M-45 and 0.3% Bavistin prior to sowing.

FYM	20-25tons / ha.
Nitrogen	100kg. /ha.
Phosphorus	75kg / ha.
Potash	100kg / ha.

Half the Nitrogen, Phosphorus and Potash are given at the time of sowing and rest 50% 50 to 60 days after sowing.

Soil and cultivation

Same as *Curcuma longa.*

Mulching

Mulching of beds with green leaves is an important and essential operation for ginger. Soon after planting, mulching with about 125 tons of green leaves / ha. Should be done. During the second and Third month of planting, mulching with 5 tons of green leaves for each mulch should be done.

Harvest

The crop is harvested after seven to eight months of sowing, when the leaves shrival, yellow and wither. The plants are forked up or dug out using spade. Care should be taken so that the epidermis is not injured as it spoils the colour of the rhizome.

Yield

100 to 150 Quintals rhizome / ha.

Curing

The Indian ginger is marketed both in scraped and unscraped form. In scraped ginger, the epidermal layer is scraped off with a sharp knife very carefully, as the essential oil remains in the epidermal cells, and careless scraping may result in loss of oil and deteriorated quality of the spice.

Bleaching

Bleached rhizomes are more in demand for export purpose in middle –east countries. For this the raw rhizomes are soaked in water for a day and later in thick milk of lime (1kg. slaked lime dissolved in 120kg. of water). The material is dried in the sun and then rubbed with gunny bag pieces to remove the skin.

In some places the lime treated rhizomes are treated with Sulphur – dioxide fumes $ 3.2kg. Sulphur per ton of rhizomes for 12 hours and then sun dried for 8-12 days.

DISEASES MANAGEMENT IN CULTIVATED MEDICINAL AND AROMATIC PLANTS

B.N. SHUKLA
Professor and Head
Dett. of Plant Pathology, JNKVV, Jabalpur

Medicinal plants are in great demand by industry therefore, have been brought under commercial cultivation. India is one of the major sources of this raw material to the world market. The cultivation of these crops has been threatened by a number of diseases, which lead to substantial losses. The present paper gives a brief account of important diseases of medicinal and aromatic crops and their method of management.

1. *Cymbopogon spp.*

Red leaf spot (*Colletotrichum graminicola*)

The leaf spot is prevalent on all the Palmarosa growing region of the country and appears during rainy season. Under favourable conditions for the growth of pathogen, the spread of pathogen is rapid leading to approximately 35 per cent reduction in the oil production. The affected parts ultimately dry out and do not yield the characteristic oil.

The pathogen perpetuates on the stumps and stubbles on the field. Therefore, under rainfed cultivation, when favourable conditions for the disease arrive during rainy season, the pathogen becomes active and spreads on the lower leaves forming small brownish spots, scattered all over the ventral surface of the leaves. Warm and humid condition favours the disease, which form enormous black fruiting bodies of the pathogen on older lesions. The disease also infects the leaf sheath and midrib. Later, many such spots coalesce to from a bigger patch. The spots show concentric ring in the centre.

Two spraying of Bavistin (0.1%) just after the first appearance of the disease at an interval of 20 days or three sprays of Dithane M-45 (0.3%) at an interval of 10-12 days check the spread of this disease.

Leaf blight (*Curvularia sp.*)

The disease is caused by *Curvularia andropogonis*, which also cause the leaf blight of *Java citronella* and lemon grass (*C. flexuosus*) in different parts of India during rainy season.

The disease appears as minute, circular, reddish brown spots, mostly on the leaf margin and tips which later enlarge and coalesce to form longitudinal, elongated, reddish brown, necrotic lesions resulting in premature drying of leaves. Older leaves are much susceptible to infection. The disease causes more than 31% loss of total oil and 17.8% of geraniol in Palmarosa. It has been reported that *C. andropogonis* appears in nature as two strains. One causes leaf spot (strain CLS) and another leaf blight (strain CLB); the strain- CLS infects all the four species of *Cymbopogon* severely whereas strain CLB is severe on *C. witerianus* and *C. flexuosus* but does not infect *C. martini* var. motia.

A leaf spot disease caused by *C. trifolii* has also been recorded to be severe on lemon grass at Central Institute of Medicinal and Aromatic Plants, Lucknow. Pathogenicity studies revealed that Curvularia causing leaf blotch on *C. martini* var, *maotia* can infect *C. flexuosus*, *C. pendulus* and *C. winterianus*. The disease is characterized by appearance of minute, circular, reddish brown pots, mostly on the leaf margin and tips which later enlarge and coalesce to form longitudinal, elongated, reddish brown, necrotic lesions resulting in premature drying of affected leaves. Older leaves are more susceptible to infection. Lenticular spots have a distinct dark brown border. The size of the spots varies from 1-8 mm in length and 0.53 mm in width. The spots often coalesce to from blotches.

Eye spot/ leaf spot (*Helminthosporium sacchari*)

The disease is characterized by oval shaped, brown spots with pale centre and dark purple borders which in advance stages coalesce to cause necrotic blight areas.

Rust (*Puccinis nakanishikii*)

Rust disease on lemon grass has been reported from several parts of the world and the fungus responsible for causing the rust is *Puccinia nakanishikii* and *P. purpurea.* Infected leaves show discrete, linear, brown uredia on the abaxial surfaces associated with chlorotic streaks, andsevere infection cause extensive necrosis corresponding with coalesced uredia. Telia also occur on severely infected leaves. No collateral hosts are known. Spraying with Dithane Z-78 (0.3%) at an interval of 10-12 days starting from July can control the disease.

Floral (*Tolyposporium christensenii)*

This disease was reported to occur severely at Indore on Palmarosa and in Kerala on lemon grass. Since the disease is inflorescence born and therefore adversely reduces the oil yield and seed production. The symptoms are characterized by the transformation of seed into the smut sori, which are slightly long covered by dull yellow membrane. Treating the seeds with thiram (0.3%) before sowing in the nursery; and spray of Dithane Z-78 (0.3%0 just at flower initiation is effective.

Root diseases

Wilt (*Fusarium moniliformae)*

The disease is characterized by pale yellow spots and wilting of the plants. A clump rot disease of lemon grass caused by *Fusarium equisetti* was recorded at the Aromatic and Medicinal Plant Research Station, Odakkali (Kerala). The disease is not much prevalent in the Cymbopogon growing regions.

2. Vetiver

Vetiver (Khus) oil is very well known for its fragrance and pleasing smell since ancient times in India. It is grown in many parts of the country like Rajasthan, U.P., Tamil Nadu, Karnataka and Andhra Pradesh.

Leaf disease

Leaf Blight (*Curvularia trifolii* Kauf. Boef.)

A leaf disease caused by *C. trifolii* was reported from India and is fairly common. Tan to dark, oval spots appear on the leaves, which later turn black as the leaves turn pale yellow. The spots become necrotic and coalesce to form blighted appearance.

Gloeocerospora leaf spot (*Gloeocerospora sorghi*)

It's a relatively new disease on Vetiver and is caused by ***Gloeocerospora sorghi.*** The disease shows small, diffused brown spots with irregular margins on leaves.

Helminthosporium leaf spot (*Helminthosporium* sp.)

The diseased spots are scattered throughout the leaf and sheath. They are oblong or oval and have a dark purplish brown colour.

Foliar disease

Smut (*Ustilago vetiveriae*)

The pathogen infects the inflorescence part of *Vetiveria zizaniodes* and in place of seeds, the smut sori are present. Spraying of Dithane Z-78, copper oxychloride or Dithane M-45 at 3 g/litre of water at an interval of 10-12 days has been found effective.

3. Mentha

A number of diseases attack this crop and cause a heavy toll. Some important diseases on Japanese mint (*Mentha arvensis* var. *piperescens*) are discussed here.

Foliar diseases

Powdery Mildew (*Eryssiphe cichoracearum*)

During April-May, when the crop demands high rate of irrigation, the warm and moist conditions favor powdery mildew, which causes heavy reduction in oil (20 per cent) due to defoliation. The disease is caused by *Erysiphe cichoracearum.* The disease appears as small chlorotic spots on the ventral surface of the leaves. Brownish discoloration is accompanied on the corresponding dorsal surface. Later on, a powdery mass appears which leads to formation of powdery patches on upper on stems and branches. Spraying of 0.3 per cent wettable sulphur or 0.05 per cent karathane is effective in minimizing the losses.

Leaf spot (*Curvularia lunata)*

This is caused by *Curvularia lunata.* Initially, minute dirty brownish spots scattered all over the leaves appear. Later when the disease advances, these develop in size and turn into big spherical or irregular patches. The symptoms of the disease may appear at any stage of the growth of the plant but bigger spots only appear on lower leaves. The remaining portions of the infected leaves become chlorotic and finally wither away.

Leaf blight (*Alternaria* sp.)

Alternaria sp. is the causal pathogen producing round to oval or slightly irregular dark brown spot on the ventral surface of the leaves. These spots consist of concentric zones, which are surrounded by pale yellow margins. In advanced stage, these spots coalesce, enlarge and occupy the entire leaf or portion of the leaf leading to defoliation.

Rust (*puccinia menthae)*

It is the most important foliar disease of mint in U.S.A. and Japan. Almost all the Mentha species are reported to be affected by rust. The disease was reported from Jammu (India) but is absent, by and large in other parts of India.

Heavy defoliation occurs which ultimately reduces the oil yield. The rust caused by *Puccinia menthae* Pers. is an authocious long cyclic disease.

Wilt (*Verticillium alboatrum*)

Wilt caused by *Verticillium alboatrum* is reported to cause heavy losses in Indiana State of U.S.A. The pathogen being soil borne infects the roots, which affects the normal growth and development of the plant. The main symptoms of disease are drying, unilateral development of branches besides causing etiolation of the leaves. This results in wilting and ultimately death of the plant. The disease spread through infected stolons when used as seed material. Use of disease free stolons for fresh planting, deep summer ploughing and well-drained field conditions held in minimizing the incidence of disease.

Stolon Rot or Stolon Decay (*Rhizoctonia bataticola*)

The disease is prevalent on the *Rhizoctonia bataticola* infected mint field. It was first reported on *M. cordiaca* and *on Mentha arvensis* and *Mentha spicata.* The initial symptoms of the disease appear as stunted growth of the plant. The leaves turn yellow. As the disease advances the over-ground portion of the plant, dies Pinkish brown lesions in the earlier stage of infection are found on stolons which gradually turn into dark brown to black patches. As the patches increase in size, soft decay of a portion or decay of entire affected stolon occurs. The recommended preventing measures are the use of disease free planting material, deep summer ploughing, crop rotation, and proper drainage of the field.

Sclerotium Rot (*Sclerotium rolfsii*)

This disease affects all the grown species of *Mentha.* White mycelia strands develop around the infected collar portion of the plant resulting in yellowing of the plant and ultimately showing wilting symptoms. In advance stages, white mustard like sclerotial bodies develop around the collar portion. The plant topples down at the collar portion and the whole plant dies. The control measures are deep summer cultivation, proper drainage, drenching of Blitox-50 (0.3%) around the infected plants and burying the sclerotia in soil.

4. Belladonna (*Atropa accuminata*)

Disease

Damping-off of seedlings is caused by *Pythium* spp. for which drenching of the field by Dithane M-45 (0.1%) is recommended. *Cercospora atropae* cause leaf spot in belladonna and forms rounded or angular spots on the upper surface of leaves, while brownish to roasted spots were restricted to leaf lamina. Blitox (0.1%) controlled this disease effectively when sprayed six times at 15 days interval.

5. Foxglove (*Digitalis purpurea and D. lanata*)

Disease

The leaf blight caused by *Alternaria* spp. is recognized by formation of brown spots on the leaves, which enlarge in size covering a larger part of the leaf. Spraying with any copper fungicide at the rate of 0.1-0.2 per cent is found effective in controlling this diseases.

6. Senna (*Cassia angustifolia*)

Disease

The crop in north-western India is affected by damping-off disease at seedling stage. At a later stage, dry rot caused by *Macrophomina phaseoli* also occurs sometimes. As a control measure, the field may be drenched with Blitox-50 (0.3%) but this provides only a partial success. Leaf spot caused by *Phyllosticta* spp. and *Cercospora* spp. are recorded on this crop, mainly during cloudy weather. These pathogens can be controlled by spraying of Dithane M-45 (0.25%).

7. Medicinal yams (*Dioscorea spp.*)

Disease

Fungal diseases caused by *Cercospora, Ramularia* and late blight by *Phytophthora infestans* reported in *D. camposita* from Tamil Nadu can be controlled by Dithane Z-78 (0.3%) and any copper fungicides spray. Collar rot of seedlings in nursery can be effectively controlled by draching of Biltax-50 (0.3%).

8. Safed musli (*Chlorophytum spp.)*

Disease

In storage, fleshy roots were found to become hollow and infection of *Aspergillus* species and *Fusarium* species were observed. The treatment of fleshy roots with thiram and captan at 4.0 g/kg reduced rotting of fleshy roots in storage and also enhanced its sprouting percentage. Collar rot caused by S. *rolfsii* is also an important disease of the crop in field for which no study has yet been conducted.

9. Sarpagandha (*Rauvolfia serpentine)*

Disease

A large number of pathogens are describes which includes a number of leaf spot diseases on Rauvolfia, such as *Cercospora* leaf spot (*Cercospora rauvolfiae)*, target leaf spot (Corynespora *cassiicola*), leaf blotch (*Cercospora serpentine*), Alternaria leaf blight (Alternaria tenuis). Besides these, Anthracnose (*Collectotrichum gloeosporioides*), die-back (Colletotrichum *demoatium*), Powdery mildew (*Laveillula taurica*) and Fusarium wilt (*Fusarium oxysporum* F. spp. *rauvolfiae*).

10. Guggal (*Commiphora wightii*)

Disease

Cercospora leaf spot has been recoded on all the cultures.

11. Opium poppy (*Papaver somniferum*)

Disease

Opium poppy is susceptible to several diseases but root rot, downy mildew, powdery mildew, sclerotinia rot and blight are some of the important diseases. Recommended doses for the control of disease of other crops have been found effective.

Root rot (*Fusarium semtectum*)

This is a very serious disease of seedlings stage and characterized by rotting underground roots. The leaves become yellow and ultimately plants die. This disease can be controlled by removing infected plants and to further check the infection, spraying of bavistin (1 g/L of water) is very effective.

Downy mildew (*Pernospora arborescens*)

This is a serious and widespread disease of opium poppy. The disease is characterized by the appearance of brownish-black coating of conidiospores and conidia of the underside of leaves. The affected leaves shrink and plants become weak and stunted. The formation of capsule is adversely affected and opium yield is significantly reduced. Since the disease is seed borne, seed should be treated with apron (EC35) or Dithane M-45 (4g/kg seed) before sowing. Spraying of seedbed with bavistin (o.o5%), or any copper fungicides (0.4%) is useful to control the disease. Spraying of bavistin (o.o5%), or Dithane M-45 (3 g/L water) at an interval of 15 days may check the infection in standing crop. Proper crop rotation and destruction of infected plants in the field are prophylactic measures.

Powdery mildew (*Erysiphe polygoni*)

It is a minor disease recorded from many localities but causes serious damage. It is characterized by white powder on leaves and capsules. To control this disease, field sanitation is important. Dusting of 'Sulphur dust' at the rate 25 kg/ha or spraying bavistin (o.o5%) is recommended to control the disease.

MANAGEMENT OF INSECT PESTS OF MEDICINAL AND AROMATIC PLANTS

O.P.DUBEY

Joint Director Extension, D.E.S., JNKVV, Jabalpur

Insects are the dominant group of animals on the earth, influencing the bio-diversity and cause enormous lossess in crop plants. Of all activities in the crop husbandry, crop protection is an important component. Hence the prevension of crop losses from pest attack is necessary to realize the high yield potential of the new varieties. No one part of the plants is free from insect pests incidence and with an estimate about 10 percent of total agricultural production is damaged by insect pests annually. Various factors are responsible for the damage to the crops and the same may also be applicable in case of medicinal and aromatic plants. Now a days medicinal and aromatic crops are of great economic importance. These crops are also associated to a number of insect pests in one or other parts of our country depending upon the agro climatic zones, season and plant type etc.

Considering the above facts an intelligent use of management tactics for different pests, diseases, nematodes and weeds in different farming situations is essential. Here we are discussing the most important pests of medicinal and aromatic plants and possibility of their management.

The insect pests of medicinal and aromatic plants are categorized in to four categories:

1. Surface and subterranean feeders Termite cutworns, white grub etc.
   (Soil insects)

2. Foliage Feeders Lepidopeterous Caterpillar, Coleopterous bettles, Orthopteron grasshoppers, crickets etc.
3. Internal feeder Various root, shoot, stem and fruit borers, fruit flies and various gall makers,
4. *Sap suckers:* Homopterous insects like aphids, jassids, white flies, sap or feeders, mealy bugs, scale insects and various heteropterous bugs viz. red cotton bug and various thysanopterous thrips.

Besides the above some of the insects are also responsible to spread plant diseases.

1. *Surface and subterranean feeders:* In this group different stages of the insect pests are residing inside the soil and they damage the plants by cutting the roots/stem portion attached to ground and due to which the plant do not get supply of food and die.

A. *Termite:* Nymphs and adults of worker caste of the termite attack young seedling as well as grownup plants; the attacked plants wither and ultimately die. Neem, neelgiri, mulethi, gulab japani pudina and lemon grass, mentha, geranium etc are attacked by this pest.

B. *Cutworm:* The slippery caterpillars are brown black in color and they cut seedling at the ground level and rood of Japani and bilayati pudiana etc.

C. *White grub:* The thick caterpillars are ofmilky white and light yellow in colour and 'C' shaped. They cut the roots of khus, chameli, isabgol etc.

D. *Beetles:* Grub of red pumpkin beetle roots damage the roots and by boring in to the under ground stem and some times in to the fruits touching the ground of sarpgandha, ashwaganda and japani pudina etc.

E. *Tuber caterpillar and root weevil:* The legless tuber caterpillars cut the tuber of ginger and onion, While root weevil feed on roots of Japanese pudina.

2. *External feeders:* The lepidopterous larvae, coleopterous grubs and adults and orthopteron hoppers and crickets are comes under this group. The different stages of these insect pests damages the different plant parts with the help of their biting and chewing type of match parts.

(a) *Surface and kharif grasshoppers:* The damage of these insect pests are found from medium to sevier in akarkara, adusa, amaltas, arjun, anantmool, ashwagandha, bel, bamboo, bahera, bramhi, safedmusli, sadabahar, satawar, shivnag, shikakai, karonda, kali haldi, keokand, marorphali, mulathi, muskdana, pudina, etc. while bach, haldi, hadjor, gataran, kalmegh kanear, ramdatoon, khus landipeeper, sitaphal, ratanjot, pamarosagrass are found resistant against this group of insect pests.

(b) *Citrus butter larvae:* The full grown larva is yellowish green with a thorn like structure on the last dorsal segment . Caterpillars feed from the margin in wards, reaching the midribs of bapchi leaves. In severe case 15-20 larvae were found feeding on babchi plants completely defoliated.

(c) *Chirak looper:* Long thin caterpillar of red and black in colour, voraciously feed on the foliage of chitrak and plants become leafless. The pest is active from July to November.

(d) *Aswagandha Caterpillar:* The caterpillar is green with yellow in colour. Due to voracious feeding of this caterpillar plants become leafless.

(e) *Hairy Caterpillars:* The red, yellow and black colour hairy caterpillars gregarious in nature and feed the green matter of leaves leaving behind only veins of arandi, muskdana, Japaneese pudina and cardamom etc.

(f) *Blister beetle:* Are of 2.5 cm long and 1 cm broad with black, yellow strips on the wings. The beetle is very dangerous to almost all the medicinal and aromatic plants but mostly found feeding on the flowers and flower buds of Jason, rose, gumchi, sada suhaganm and aprajita. Two to three beetles flower were almost found feeding on Jason and rose plants.

(g) *Haldi leaf folder:* The long, thick larva of green in colour with black head, spin the leaf of haldi, kali haldi and amahaldi and feed inside. About 5-10 per cent plant were found damage by this pest.

(h) *Epilecna beetle:* This beetle is round in shape with black circuler markings on wings. The adult and larva both are feeding on the leaves of janglibhata and bhatkatai

3. *Internal feeders:* The mouth parts of the insect pests belonging to this group are of biting and chewing type but the larva feed internally.

The number of internal feeders are small enough to find comfortable quarter.

(i) *Leaf minor:* The larvae of this group mines the leaves of jamun, bahera, muskdana, arandi khamer, umar and gataran. Out of these medicinal plants the damage was found more in gataran, khamir, muskdana and arandi plants.

(ii) *Stemborer:* The larva of stemborer feeding inside the main shoot of khas, cardamom, citronella, aonla and muskdana, The attack of the pest always confind to the top of the plant. The central shoot when tunneled and damage by borer, dries up and giving rise to "dead heart"

(iii) *Fruit borers:* The damage of different borers are caused in the fruits of anar, muskdana, opium, bhatkatai, khurasani ajwine, kimach etc.

4. *Sap suckers:* The nymphs and adults of hemipteron and thysan- epteron, suck the sap from different parts by their piercing and sucking type of mouth parts. Due to the damage of these pests' plants become weak,

curling of leaves, browning and dying of the whole plants.

A. *Aphids:* Both nymphs and adults suck the sap mostly from the apical leaves and tender twigs of opium, ashwagandha ajwine, Japanese and vilayati pudina, isabgol, jason, muskdana, kapoor washi, saunf, coriander, rose, jasmine, mentha, and belladona etc.

B. *Jassids choppers:* The nymphs and adults suck the cell sap from different tender parts of safed musli, kapoorbasi, chandan, vilayati pudina etc.

C. *White files:* The incidence of this pest is mostly found in muskdana, mentha, lemongrass, vilayati and japaneese pudina and bhatkataj, Due to attack of this pest infested leaves become clonotic and eventually wither. The pest sucks the sap and cluster on the lower surface of leaves. The damage is also done by this pest by secreting honeydew which interfere with the photosynthesis of the plant due to which shooty mould develop. Thus normal photosynthesis of plant is adversely affected. They have remarkable ability to transmit plant diseases. In sever out break, the growth and development of the plants is checked considerably.

D. *Mealy bugs:* The nymphs and adults of these pests suck the sap from mango, anar, karonda, sweetneem, ashwagandha, sitaphal etc.

E. *Scale Insect:* The hemispherical scales is very much convex, smooth and brown they rarely move from the plant where that young nymphs settled and insert their mouthparts. They suck the sap of plant and growth of the plant is checked and vigour is reduced, ultimately the plant dies. Thay are mostly seen on mango, vetiver, sarpgandha, rose, lemon grass etc.

F. *Thrips :* Damage to the floral parts of the palmarosa and isabgol are done by the nymphs and adults of thrips, mostly in the month of October-November in Northern parts of India. Due to serious attack of this pest seed setting arises in palmarosa.

G. *Bugs:* different type of bugs are also feeding on medicinal and aromatic plants. Red cotton bug damage the fruits of muskdana by sucking the sap.

Integrated Pest Management (IPM)

Integrated pest management (IPM) is their selection, integration and implementation of pest control based on predicted economics and ecological and sociological consequences. The Government of India has adopted IPM as a major thrust area and the main plan of crop protection programme. IPM approach is eco-friendly. It aims at minimal use of hazardous chemical pesticides and instead employing alternate pest control methods and technique like cultural, mechanical biological and biopesticides.

Soil conditioning, crop rotation and improved sanitation practices are some simple methods to reduce pest problems. The crops can also be protected by eliminating pests by methods of starvations, where a target crop is grown to lure insects away from the economic crop. Sometimes, to obtain desirable pest control, biological methods are supplemented with chemical pesticides. Employing the natural enemies of crop pest like parasites and predators and a variety of bio-control agents, such as beneficial bacteria and viruses, have been successful in the management of pests. Among the cases of IPM success are the package for staple crops viz. sugarcane, cotton, citrus, vegetable etc.

Surface or subsurface feeders (soil Insects)

1. Clean cultivation:- Keeping the fields and bunds free from weeds, destruction of first attacked plants or plant parts, timely disposal of plant residues after harvest /pickings.

2. Deep plouging after removal of crop at least after 3years.
3. Trapping of cutworm caterpillars in heaps of refuse and moths in light trap.
4. Putting bands of insecticides dust around the cutworm population.
5. Judious and timely irrigation: Water may be charged with crude oil emulsion or thin film of kerosene.
6. Use well decomposedr organic matter.
7. Avoid greenmannuring and raw farmyard manure.

Chemical

1. For termite control drenching of chlorpyriphos 20 EC or endosulfan 35 EC @ 3-5 Lit/ha or dusting of lindane 1.3% @ 40 kg/ha at the time of field preparation or planting.
2. Treat the trees with chlorpyriphos 20 EC 3-5 ml/liter of water for killing the adult beetle.
3. Soil application of phorate 10 G @ 10 kg/ha. at the time of field preparation/ planting.

External feeders

1. Removal of egg masses or clusters of larvae and pupae on leaves and their destruction regularly very effective against lepidopterous hairy caterpillars and colepterous beetles (epilachna)
2. Timely planting/sowing of medicinal and aromatics reduce the infestation of various pests.
3. Use of light trap to collect the noctuid, and arctid moths.
4. Locust adults may be killed by using flame throwers.
5. Trenches may be dug on way of marching hoppers and after falling in them may be burried alive.
6. Plough of the field after harvest to kill the pupae and grub of redpumkin beetle and egg masses of grasshoppers.

Chemical

1. Spray the crop with chlorpyriphos 20 EC or endosulfan 35 EC @ 2ml/liter of water or Malathion 50 EC or quinolphos 25 EC @ 1 ml/lit of water. Repeat the same if necessary.

Internal feeders

1. Removal of drooping shoots, bored fruits and mined leaves and their destruction.
2. Controlling the pests on various alternate host plants.
3. Fruit fly seeking shelter under leaves in sunny weather should be killed by spraying of chemicals.
4. For leafroller timely sowing is important.
5. Citus leaf minor can be controlled by prunning of heavily affected plant parts during winter and burn them.

Chemical

Spray the crop with chlorpyriphos 20 EC or endosulfan 35 EC @ 2 ml/liter of water or malathion 50 EC or quinolphos 25 EC @ 1 ml/liter of water.

Sap feeders (sucking insects)

1. Early sowing can minimize the incidence of aphids.
2. Close planting, water logging and other trees should be avoided (white fly).
3. Removal of affected parts.
4. Plough or dug repeatedly around the tree and apply insecticides (mealy bugs)
5. Use grease bands around the tree trunk to prevent the young nymphs from climbing (mealy bug).
6. Collect fruit moths during initial stage of attack by means of hand net and bag the fruits with plastic or cloth bag.
7. Destroy wild host plants in the vicinity of orchards.

8. Bait moths with lead arsenate and gur solution mixed in proportion of 1:160 and put in small pots, hung on branches add few drop of vinegar to attack moths.

Chemical

The crop should he sprayed by with methyl demeton 25 EC or trizophos 40 EC @ 2.5 ml/lit or of water or dimethoate 30 EC @ 1 ml/liter of water and repeat if necessary.

Some useful plants/weeds can alsobe used for reducing the incidence of these pests. Generally neem, lantana, notchi, tulsi, adathoda etc. act as repellent to many pests for a shorter period.

The indigenous treed like pungam, wood apple, anona and their byproducts have excellent insecticidal value in controlling diamond back moth, heliothis, white files, leafhopper and aphid infestation. The fish oil resin soap (FOS) is a safe, non-posionous natural product used for the control of developed whiteflies, mealybug, wooly aphid etc. A wider range of bio extracts are proved to be more effective in controlling many insect Pests and Diseases.

UTILIZATION OF WASTELANDS BY CULTIVATING MEDICINAL PLANTS UNDER AGROFORESTRY SYSTEM

S.D. UPADHYAYA AND AASHUTOSH SHARMA

Department of Plant Physiology, JNKVV, Jabalpur (M.P.)

Preamble

In a scenario of decreasing availability of good lands for agriculture, degradation of soil and water resources, increasing pollution of soil and water resources, increasing pollution hazards and threats to the environment and ecosystem, new approaches in farming systems are needed to meet food, fodder, fiber, firewood and timber targets fixed for the 21st century. The future agricultural technologies must assure sustainability while optimizing productivity. Deliberate inclusion of trees in the existing farming/ cropping systems and adoption of agroforestry approach on all kinds of wastelands is one of the viable options to achieve sustainability while optimizing productivity.

Significant Global Environmental Challenges For The Next Decade

- √ Sustainable management and use of natural resources
- √ Sustainable production and consumption
- √ A better environment for human health and well-being
- √ Biodiversity conservation.

Arable land available for agricultural production and world population

Year	Arable land A/person	Population (in billion)
1965	1.14	3.0
1980	0.84	4.5
1990	0.74	5.5
2000	0.62	6.5
2025(projected)	0.49	8.0

FACTS

1. 1.4 billion (ha) world cropland
2. 7.4 billion (ha) grassland, forest, fallow and wasteland
3. 1.0 billion Indian population
4. 0.5 billion livestock (15% of global)
5. 0.5% world pastureland, 1% forest

Challenges facing Indian agriculture

- Food Security
- Imbalanced Growth
- High Post-Harvest Losses
- Sustainability
- Low/declining Productivity
- Degradation of Natural Resources
- Raising Rural Income Through Diversification
- New Market Demands including Exports
- Nutritional Security

Wastelands

The lands, which are currently not under proper utilization and have a very poor productivity, are known as wastelands. It can be described as "degraded land" which can be brought under vegetative cover with reasonable effort, and is currently underutilized and is deteriorating for lack of appropriate soil/water management on account of natural causes. In India, wastelands constitute a sizeable part of the total geographical area. It has been estimated that more than half the geographical area i. e. 175 million hectares is suffering from some form of degradation, such lands are reported to as wastelands. There is a great risk of growing food grains in extremely light textured soils and culturable wastelands. This particular area could be brought under cultivation of medicinal plants.

Agroforestry – A concept

Agroforestry a term coined in 1977 denotes an age-old practice of having trees mixed in the agricultural landscape. It designates a holistic approach to land use in which woody perennial are deliberately grown on the same land management or in temporal sequence. There must be significant ecological and economical interactions between woody and non-woody components. The objective of such combinations is to increase, sustain and diversify the production of land thereby to help in reducing economic and environmental risks.

Agroforestry can be seen as phases in the development of a productive agro ecosystem. Within this ecological framework, farmers can manipulate and manage their trees to take advantage of the benefits or the processes in ecosystem services or products. Therefore, it is suggested that agroforestry should be reconsidered as a dynamic, ecologically based, natural resource management system that, through the integration of trees in farm and rangeland, diversifies and sustains smallholder production for increased social, economic and environmental benefits.

Advantages of Agroforestry

The inclusion of leguminous trees on poor soil can result in marked improvement in soil fertility by -

- Increasing soil organic matter through addition of leaf litter and plant debris
- Efficient nutrient cycling

- Efficient biological nitrogen fixation (Approx. 100-500 kg N/ha/year)
- Little loss of nutrients from the system (with holding nutrients that can be lost by leaching)
- Little erosion run off
- Additional nutrient economy (uptake from deeper layers and deposition on surface layer via litter)
- Nutrient released by root decay
- Increased activity of favourable microorganism.

Improvement in soil physical conditions such as water holding capacity, aggregate stability, soil temperature regimes, permeability, drainage etc. by

- Adding shoot/root biomass, about 3000 kg dry matter/hectare / year.
- Nitrogen supplement from Nitrogen Fixing Tree species (NFTs)

Wastelands in Madhya Pradesh

After the formation of Chhatisagarh, the geographical area of Madhya Pradesh has reduced to 307.44 lac ha., out of which 49 per cent is under cultivation. The majority of cultivated area is under rainfed (69%) with enormous unexploited agricultural production potential. Nine distinct agroclimatic zones of the State are endowed with large variations in rain fall (1800 - 1600mm), irrigation potential (3 - 64%), cropped area (3.5 to 33 %), more than 60% is non cropped; soil types (alluvial) terrain from plain to high mountain, one third of the net sown area is unutilized culturable wastelands (having less than 50% production). These situations pose threat to crop yields and warrant for an alternate land husbandry and crop diversification for sustainable production. There is a great risk of growing food grains in extremely light textured soils and culturable wastelands of Madhya Pradesh, which occupies 22% of the total geographical area of the State. This particular area could be brought under cultivation of medicinal plants.

According to agricultural statistics (1993) the undivided Madhya Pradesh has 1.68 M. hectare area under culturable wastelands. Beside this, the State accounts for a large percentage of land under pastures and grazing lands. The total wasteland area of the divided state (after formation of Chhattisgarh) is 19.3 percent of the geographical area includes major portion under gullied/ravinous land, degraded notified forestland, and barren rocky areas.

Categories of Wastelands in the State

The wasteland percentage in the districts to total geographical area of the State varies from 6% (in Balaghat) to 41% (in Tikamgarh). There are ten major categories of wastelands in the State namely,

1. Gullied/ravinous land
2. Land with/without scrub
3. Waterlogged/ Marshi land
4. Saline/ Alkaline Area
5. Deg. Notified forest land
6. Deg. Pastures/ grazing land
7. Deg. Land under plantation crop
8. Mining/ Industrial wasteland
9. Barren rocky area
10. Steep slopy area

The major area belongs to gullied/ ravinous and land with and without scrub.

Scope of medicinal plants cultivation

Culturable wastelands, which have the potential for the development of vegetative cover, can easily be brought under cultivation of medicinal plants through proper land preservation and species selection. The problem areas like ravines, saline and alkaline soils, canal side wetlands, seasonally inundated forests, pasturelands and agriculture lands needs special attention for their proper use and applicable technology for its development. The technique of agroforestry involving planting of medicinal trees/shrubs and herbs in wastelands or in problem area as alternate land use not only provides the solution of reclamation of wastelands but also gives products wanted by local communities. Agroforestry systems involve deliberately growing woody perennials/medicinal trees and shrubs along with arable crops and/or livestock seeking positive synergism. Traditionally several species of trees having medicinal values e.g. Neem, Babool, Harra, Bahera, Aonla were grown in wastelands. Farmers are growing these trees on farm land for different uses involving fodder, fuel, timber, gum, medicine, oil, fruit, bee keeping, green manure etc. Further, these woody perennials also help in soil amelioration. They not only add carbon to the soil but also enrich it with nitrogen and assist in nutrient retrieval from deeper layers for final use by arable crops. The soil physical conditions also would be ameliorated. They act as windbreaks and help in living with the problems like acidity, alkalinity, and water logging. Between medicinal trees, there will be differential addition of nutrients and the soil will have a regeneration of microflora depending on the tree species. Medicinal plants can also be grouped as suitable to different edaphic factors like shallow soils, rocky terrain and sandy soils (table 1). They also can be identified for different soil problems like salinity, sodicity and waterlogging (table 1).

Inter cropping of medicinal herbs with various tree species

Farmer's main aim of cultivating wastelands is to evolve a system that has positive synergism. It is practically axiomatic that trees/woody perennials have potential benefits in soil amelioration in addition to productive functions as medicinal values, such as:

(a) Enrichment of carbon and nitrogen
(b) Reduction in soil erosion and nutrient loss
(c) Nutrient uptake from lower layers to the surface
(d) Increased microbial activity
(e) Improvement in litter quality, and
(f) Work as shelter belt/ wind brakes.

In wastelands, the suitable tree and herb species can be selected and grown together, generally two years before the tree canopy develops to cover the interspaces fully. In this system, the medicinal trees are primary crops and healing herbs are secondary. Inter cropping can be taken for variable periods. Some of the medicinal herbs requires shade, for which close planting trees with good coverage of canopy can be preferred; such systems are known as shaded perennial systems, for instance; Aonla with turmeric or Khamer with Chitrak.

Conclusion

Wastelands literally means the land which is uncultivated, barren or without vegetation. The areas those for any reason are presently unproductive or productive at a level very much below their potential. The soil properties most often associated with wastelands are steep slope, lack of fertility, salinity, acidity, stoniness, shallow soils, eroded soils, wetness or flooding. These are otherwise known as problem soils. Farmers have cultivable wastelands which has some soil but not much productive. Such cultivable wastelands may be turned productive by selecting especially tolerant species and varieties of arable crops/trees. Medicinal plants are best ameliorators in living with problem soils. Most of the soils can be reclaimed by planting trees. Class V andVI lands belongs to enterprising farmers/lands held by small communities should be brought under cultivation by encouraging planting of medicinal plants/trees through participatory approach.

Table 1: Suitable medicinal plants that do well on different problematic soils.

No.	Problematic Soils	Name of the species	
A	Shallow, rocky soil	**Herbs**	Punarnava (*Boerhavia diffusa*) Anantmool (*Hemidesmus indicus*), Ratti (*Abrus pricatorius*), Manjistha (*Rubia Cordifolia*), Gwar patha (*Aloe barbadensis, Aloe vera*).
		Shrub	Gandh Babool (*Acacia farnisiana*), Jhand (*Prosopis cineraria*), Karonda (*Carissa carandus*) Gurmar (*Gymnema sylvestre*) Marorphali (*Helicteres isora*).
		Trees	Neem (*Azadirachta indica*), Aonla (*Emblica officinalis*), Sitaphal (*Annona squamosa*), Bel (*Aegle marmelos*), *Cordia oblica, Holoptellia intigrifolia, Wrightia tinctoria* Malkagni (*Celastrus paniculata*).
B	Sandy soils	**Herbs**	Khus (*Vetiveria zizanioides*), *Cymbopogon* spp., Isabgol (*Plantago ovata*), Safed musli (*Chlorophytum borivilianum, C. tuberosum*),Kali Musli (*Curculigo orchioides*), Asgandh (*Withania somnifera*), Mulethi (*Glycyrrhiza glabra*),Senna (*Cassia angustifolia*), Akarkara (*Spilanthus acmela*), Sadasuhagan (*Catharanthus roseus*).
		Shrub	Chitrak (*Plumbago zeylanica*), Sinduri (*Bixa orellana*)
		Trees	Bel (*Aegle marmelos*), Katha (*Acacia catechu*), Jamun (*Syzygium cumini*), Arjun (*Terminalia arjuna*), Salai (*Boswellia serrata*), Khamer (*Gmelina arborea*), *Glyricidia sapium*.
C	Saline soils	**Herbs**	Satawar (*Asparagus racemosus*), Sankh puspi (*Evolvulus alsinoides*),

No.	Problematic Soils		Name of the species
			Khus (*Vetiveria zizanioides*), *Cymbopogon* spp., Chandrsur (*Lepidium sativum*), Kalongi (*Nigella sativa*)
		Shrubs	Karonda (*Carissa carandus*), Adusa (*Adhatoda vasica*), Nirgundi (*Vitex negundo*), Ber (*Zizyphus* spp.
		Trees	Neem (*Azadirachta indica*), Aonla (*Emblica officinalis*), Katha (*Acacia catechu*), Arjun (*Terminalia arjuna*), Babool (*Acacia nilotica*), Bahera (*Terminalia bellerica*), Harra (*Terminalia chebula*), Sivnag (*Oroxylum indicum*).
D	Alkaline soils	**Herbs**	Sankh puspi (*Evolvulus alsinoides*), Khus (*Vetiveria zizanioides*), *Cymbopogon* spp., Jangali Pyaj (*Urginia indica*), Van lahsun (*Allium porrum*), Garlic (*Allium sativum*).
		Shrubs	Karonda (*Carissa carandus*), Nirgundi (*Vitex negundo*), Guggal (*Commiphora* spp.)
		Trees	Aonla (*Emblica officinalis*), Arjun (*Terminalia arjuna*), Babool (*Acacia nilotica*), Sissoo (*Dalbergia sissoo*), *Eucalyptus teritecornis*, Siris (*Albizzia procera*), Karanj (*Pongamia pinnata*)
E	Water logged soils	**Herbs**	Bach (*Acorus calamus*), Brahmi (*Bacopa monniera*), Mandook Parni (*Cintela asiatica*), Brangraj (*Eclipta alba*), *Cyperus* spp. Khus (*Vetiveria zizanioides*), Talmakhana (*Astercantha longifolia*).
		Shrub	*Hedychium spicatum*
		Trees	*Eucalyptus teritecornis*, Arjun (*Terminalia arjuna*), Salai (*Boswellia serrata*)

MEDICINAL AND AROMATIC PLANTS FOR PLANTING IN CULTURABLE WASTE LAND AND FALLOW LANDS

J.P. TIWARI

Rtd. Dean, College of Horticulture Mandsaur, JNKVV, Jabalpur

After the food security due to "Green Revolution", there is urgent need of nutrition, environment and health security. At present there is acute shortage of most of the raw drug material for preparation of the medicines used in Indian Systems of Medicines. The storage of these herbs will further increase in this millennium in our country and other parts of the world owing to the increasing population as well as awareness regarding the herbal medicines, the demand of medicinal and aromatic plants raw drug material is increasing at the rate of 10 percent per year. However, the supply from the natural forests is reducing due to over exploitation and negligible efforts for the regeneration of useful herbs. Some of the species are endangered and are at the verge of extinction e.g. *Gloriosa superba, Rauvolfia serpentina, Bacopa monnieri, Commiphora wightii, Saraca asoca, Chlorophytum borivilliamum, Oroxyum indicum, Dioscoria deltoidea, Stereospermum sauveloens, Swaratia chiraa, Curcuma caesia, Curcuma angustifolia, Acorus calmus, Adensonia digitata* etc. Many species are much threatened, If the existing process of exploitation of herbs will remain contunes, many more biodiversity of usefull medicinal and aromatic plants will not be available within few years. The short supply has encouraged the use of alternatives and adulteration, consequently the authenticity of the drugs is questioned. This will reflect on the health of the common people, as 80% of the population is still dependent on the drugs derived from herbs. Therefore, for health security, the cultivation of these medicinal and aromatic plants on available fallow land is essential.

The growing of medicinal plants in fallow lands will also conserve the eco-friendly environment and reduce the pollution. Certain medicinal trees and herbs serve as sink of air pollutants and help reduce air pollution through scavenging. These species include *Ficus spp. Tectona grandis, Polyalthea longifolia, Thespesea populnea,* and *Nrium indicum* which are efficient dust collectors, while, *Solanum melongina, Cyamopsis tetragonoloba Albezia lebbek,* (Siris), *Azadiracnta indica* (Neem), *Casurina equisitifolia, Ficus relegiosa, Delonex regia* (Gulmohar), *Ailianthus* (Maha neem), *Alstonia scholaris* (saptparni), *Largistronia, Mimusops elengii* (Molsari), *Parkinsonia aculeata, Terminalia arjuna* (Arjun), *Polyalthia longifolia* etc absorb SO2 and reduce the pollution (Chadha and Choudhary, 2001). A 500 m wide green belt around the industrial units combats 70% pollution. *Fagus orientalis, Quercus robur, Robinia pseudocacia, Sambucus nigra, Ulnus* etc. against oxides of nitrogen, *Acer nigundo, Acer saccharinum, Acer platanoides, Quercus palustris, Q. rubra* etc against peroxyacetyl nitrate, *Ailianthus, Betula, Juniperus, Forsythia* against fluorides; *Ziziphus martiana* (Ber) and *Cassia fistula* against zinc, *Acer platanoides, A negundo, A rubrum, Quercus robur, Q rubra* etc. are ideal for reducing ozone pollution. The planting of medicinal plant in industrial yards, road and railway lines not only provide the raw drug material but also check the dust, smoke and noise pollution. *Albezia lebbek, Alstonia scholaris, Azadirachta indica, Butea monosperma, Cassia fistula, Eucalyptus citriodora, Ficus spp, Polyalthea longifolia, Terminatia spp. Pongamia pinnata, Shorea robusta, Syzygium cummini* etc, can be planted.

As athe Agricultural land is limited the cultivation of the medicinal and aromatic crops can be promoted under the wastelands and fallow lands so that these crops may not compete or replace the existing crop area and production of food grain will not be affected. Alternatively, these crops can be grown in the marginal lands where uneconomic crops are presently cultivated viz. kodon, kutki, niger etc. These may also be grown as inter crop with the orchards and field crops. The fallow/wastelands in the following areas should be considered for growing the medicinal trees, shrubs and herbs on priority basis.

1. Field bounds
2. Garzing lands
3. Undulated lands on the banks of rivers, nalas etc.
4. Water logged area near tanks, dams and canals
5. Unirrigated Hilly lands
6. Fallow land of court yards of the houses
7. Fallow land of court yards of Panchayat, Mandi, School, College and other government buildings
8. Fallow lands by the side of roads and rail lines
9. Fallow lands in factory yards.
10. Fallow lands in courtyards of temples, mosque, Gurudwara, burial places, chopal, etc.

The fulfill the future need of herbal drugs and combat the environmental pollution no land should be left barren rather be covered by planting the medicinal and aromatic trees and shrubs. This will provide health security for coming generation.

Table 1: Medicinal and aromatic trees, shrubs and herbs of different kinds for planting under wasteland.

Trees	Shrubs/climbers	Herbs
1- Field bouds		
Embelica officnalis	*Lawsonea enermis*	*Aloe vera*
Santatum album	*Annona squamosa*	*Chlorophytum borivillianum*
Eucalyptus citriodora	*Embelia ribes*	*Costus speciosa*
Aegle mar malos	*Asparagus racemosus*	*Ricinus communis*
Bucha nania lanzan	*Gymnema sylvestre*	*Cassia angustifolia*
Gmelina arborea	*Acacia concinna*	*Andrographis paniculata*
Moringa oleigera	*Gloriosa superba*	*Curcuma angustifolia*
Semicarpus anacardium	*Commiphora wightii*	*Rauvolfia serpentina*
Ziziphus jujuba	*Citrus pp.*	*Convolvulus pluricaulis*
Holarrhena antidysentrica	*Jatropha curcus*	
Dioscoria		

2. Fallow lands and Grazing lands

Azadirachta indica
Gemelina arborea
Oroxyllum indicum
Embelica officinalis
Terminalia bellirica
Terminalia chebula
Shoerea robusta
Ficus religiosa
Stercutia urens
Aegle marmelos
Limonia acidissima
Tamarindus indica
Pongamia pinnata
Eucalyptus globules
Eucalyptus citriodora
Butea monosperma
Dalbergia sisso
Albezia lebbek
Anogeisus latifolia

Tinospora cordifoloa
Mucuna prurita
Embelia ribes
Acacia concinna
Sapinds mukurosi
Ziziphus jujuba
Ocimum gartissimum
Nyctenthus arbortristis

Annona squamosa
Gymnema sylvestre
Jatroptia aurcus
Clerodendrum phlomidis
Datura alba
Mucuna pourita

Abutilon indicum

Cassia angustifolia
Tephrosia Purpuria
Cassia tora
Boerhaavia diffusa
Ocimum canum
Aloe vera
Ocimum basalicum
Tribulis terrestris

Solanm xanthocar pum

3 – Stream/river banks

Terminalia arjuna
Syzygium comuni
Eucalyptus globulus
Lamala

Vitex nigundo
Barlaria cristata
Pandanus odoratissmus
Wood fordia
Alpenia galanga

Vetiveria zizanioides
Acorus calamus
Centill asiatica
Cyperus scariosus
Hedychium spicatum

4 – Water logged area

Terminalia arjuna
Syzygium commini
Eucalyptus globulus

Pandonus odoratissimus
Nymphea nouchali
Piper longum

Acorus calamus
Bacopa monneri
Vetiveria zizaniodes
Hedychium coroharium

5 Unirrigated Hilly areas

Anogeissus latifolia
Emblica officinalis
Terminalia bellirica
Terminalia chebula
Buchanania lanzan
Semicarpus anacardium
Azadirachta indica
Sapandis mukurosi
Oroxylum indicum
Gmelina arborea
Acacia nilotica
Bombax ceiba
Bambusa arundinacea
Albizia lebbek
Pangamia pinnata

Annona squamosa
Acacia concinna
Gymnena sylvistre
Dioscorias
Helecteris isora
Tinospora conrdifolia
Bursera panicillata
Asparagus
Hemidesmus indicus
A cacia catachu
Embelia ribes
Jatropa curcus
Adhotoda vasica

Curcuma angustifolia
Curcuma amada
Curcuma caesia
Zingiber zeopordes
Terphrosia purpuria
Aloe vera
Ocimum canum
Chlorophytum boribillianum
Androyraphis paniculata

Ocimum graticimum
Boerrhavia diffusea

6 – Fallow land of courtyards of houses

Emblica officinalis
Cicca acida
Moringa oleifera
Ziziphus martiana
Santalum album
Crateava nuravala
Citrus acida
Michalia Champaca

Anona squamosa
Punica granatum
Bixa orellana
Commiphora wightii
Jasminum samboc
Tinospora cordifolia
Asparagus racemosus
Ocimum sanctum

Chlorophytum borivillianum
Aloe vera
Curcuma longa
Catharanthus roseus
Zingiber officinalis
Curcuma angustifolia
Curcuma caesia
Piper asiaticum
Bacopa monnieri

7 Fallow lands of Panchayat, Mandi, Schools, College and other government buildings.

Azadirachta indica
Cassia fistula
Delonex regia
Tabebuia spectabilis
Teminialia bellirica
Bombex ceiba
Pterocarpus marsupium
Oroxylum indicum
Stereospermum sauveolens
Saraca asoca
Michelia chmpaka
Kigelia pinnata
Cassuarina equisetifolia
Pongamia pinnata
Bauhinia variegata
Boswellia serrata
Caryota urens
Spathodia campanulata
Delonix regia
Largerstromea speciosa
Adensonia digitata
Caryota urens
Cassia glauca
Cassia grandis
Cassia javanica
Dalbergia sisso
Dandrocalamus strictus
Dillenia indica
Grevillea robusta
Hevea brasilensis

Gardinia
Nyctenthes arbotristis
Acacia concinna
Bixa orellana
Plumbago zeylanica
Clerodendrum phlomidis
Nerium indicum
Pandanus odoratismus
Thuja orientalis
Cupressus torulosa
Aspargus racemosus
Gardenia jasminoides
Bursera panicillata
Jatropa curcus
Carissa carandis
Celastrus paniculata
Cesternum nocternum
Cesternum diurnum
Clerodendrum Phlomidis
Pandanus odoratisma
Clerodendron serratum
Rosa damascina
Jasminum arborescens
Jasminum sambac
Cassia sophora
Jasminum flexile
Hibiscus rosa sinensis
Plumeria alba
Plumeria rubra

Rauvolfia serpentina
Costus speciosus
Crinum asiaticum
Chlorophytum borivillianum
Curculigo orchiodes
Evolvovulus pluricaulis
Catharanthus roseus
Acorus calamus
Coleus aromatica
Argeria indica
Curcuma amada
Curcuma caeisa
Curcuma angustifolia
Canna
Crinum defixum
Crinum latifolium
Crinum zeylanicum

8 – Temple, Gurudwara, Mosque Yards

Anthocephalus kadamba
Aegle marmelos
Bauhinia purpurea
Bauhinia variegata

Cesternum nocternum
Murraya paniculata
Cesternum diurnum
Gardenia jasminoides

Centilla asiatica
Ocimum basalicum
Ocimum graticimum

Michalia champaca
Eleocarpus sphaericus
Mesua ferra
Adansonia digitata
Polylthea longifolia
Putarnjiva roxburghii
Pterocarpus marsupium
Pterocarpus santalianus
Pterospermum acerifolium
Pterospermum canescen
Satalum album
Saraca asoca
Stereopermum suaveolens
Strycons potatrum
Strycnos nux-vomica
Ficus benglensis
Ficus krishnea
Ficus religiosa
Ficus glomerata
Tabebuia spectabilis
Tabebuia pentaphylla

Wrightia tinctoria
Cicca acida
Cinnamomum camphora
Sinnamomum zeylanica
Cinnamomum tamala
Alstonsa scholaris
Butea monsperma
Caryota urens

Cassia fistula
Cassia glauca
Crataeva nurvala
Delonix regia
Dendrocalmus structus
Dillenia indica
Oroxylum indicum
Bombax ceiba
Lagerstrmia speciosa
Gliricida muculata
Azadiracta India
Thespesia populnea
Pongamia pinnata

Nerium oleander
Nerium indicum
Pandanus odoratissmus
Pavonia odorata
Plumbago zeylanica
Plumeria alba
Plueria rubra
Rosa damscena
Bixa orellana
Argyria nurvosa
Aristolochia indica
Cycus circinalis
Tylophora indica
Tinospora cordifolia
Thuja occidentalis
Thevetia peruviana
Tecomella unduluta
Tecoma stans
Annona muricata
Taeberanae montana
Annona reticulata divaricata
Vitex nigundo
Cissus quadrangularis
Calotropis procera
Datura metel
Cissampelos pareira
Commiphora wightii
Alpenia galanga
Balisopermum montanum
Asclepias curassavica
Barlaria alba
Barlaria crustata
Berberis aristata
Bryophyllum pinnata
Clerodendron indicum
Clerodendron serratum
cupressus torulosa
Gloriosa superba
Celastrus paniculata
Jasminum arborescens
Jasminum multiflorum
Jasminum aurichulatum
Jasminum flexile
Jasminum sambac
Hibiscus rosasinensis
Hibiscus tiliaceus
Abrus precatorius

Ocimum canum

Ocumum sanctum
Lavendula angustifolia
Bacopa monnieri
Chlorophytum borivillianum
Mirabilis jalapa
Rauvolfia serpentina
Ruellia tubrosa
Salvia splendens
Sensveiria zeylanica

Cathranthus roseus
Symbopogon flexuous
Urginia indica

Tephrosia purpurea
Boerrhabia diffusa

Adiantum capillus veneris
Withania somnifera

Crinum latifolium
Acorus calamus
Destostechea bipinnata
Solanum xanthocarpum
Tribulus terrestris
Aloe vera
Cassia angustifolia
Solanum xanthocarpum

Convolvulus pluricaulis
Costus speciosus
Citrullus colocynthis

Coleus aromatica
Curculigo orchiolides
Curcuma aromatica
Curcuma caesia
Curcuma amada
Curcuma longa
Aristolochia bractiolata
Aristolochia indica
Artemisia pallens

Medicinal Trees for Planting by the sides of Roads, Railway and factory yards (Also reduce dust, smoke and noise pollution)

Acacia auriculiforma
Albeza lebbek
Alstonia scholaris
Azadirachta indica
Butea monsperma
Cassia fictula
Cassia siamea
Delonex regia
Dalburgia sisoo
Erythrina indica
Eucalyptus citriodra
Fucus benjmena
Ficus bengalensis
Ficus infectoria
Ficus infectoria
Grevillea robusta
Juniperis chinensis
Kigelia pinnata
Madhuca indica
Mangifera indica
Mimusops elengii
Nerium indicum
Nerium oleander
Peltaphorum ferrugineum
Polyalthea longifolia
Shorea fobusta
Syzygium Cuminii
Tectona grandis
Terminalia arjuna
Terminalia bellrica
Terminalia chebula
Thevetia peruviana

IMPORTANCE OF SEED GERMINATION AND VIGOR TESTING

SATHRUPA RAO
Principal Scientist, Seed Technology Research Unit,
Department of Plant Breeding and Genetics, JNKVV Jabalpur

Objectives of germination testing

The aim of testing breeds for germinability is to gain information with respect to planting value of seed. Hence germination defined by (ISTA) as the emergence and development of the seedling to a stage where the aspects of its essential structure indicates whether or not it is able to develop into a normal plant under favorable conditions. The field emergence ability is the major aspect of seed quality of concern to growers and high germination is the pre-requisite.

Seed trading

The second importance of germination test is to provide results which can be used to compare the value of differeal seed lots (ISTA 1985). The outcome of the germination test may decide

- Whether the seedlots meets the statutory or voluntary germination standards.
- Whether two or more seedlots require blending to produce a single lot of acceptable germination.
- Whether the seedlotes suitable for export and /or internal distribution and sale.
- Whether the seedlot is within the seed certification standards.

Seed storage

The initial quality of the seed determines these potential longevity under any storage conditions thus on testing the germination of a particular seedlot one can assume the storage capacity of the seed lot and can take some protective measures as monitoring and moist ring, Humidity of the store. In order to stower down the insect pest management and pathogen infestation seed deterioration.

Limitations of germination test

- It adequately meets only one of the three objectives.
- As per ISTA rules the substrate on which the seeds are tested must at all times contain sufficient moisture but the amount of moisture is not specified.
- The methods for temperature of hous only one temperature or alternating temperature should b optimal.
- During assessment of germination only normal seedlings are taken into account however there is no account of weak seedlings. In practice it is only the dead, badly diseased and irrevocably abnormal seedlings are eliminated. The weak seedlings, semi-weak and robust seedlings are given equal weight age.

Hence keeping in mind the variable performance of a seed lot during germination testing in the lgboratory and its performance in the field can help the seed analyst or the grower to eliminate a particular seedlot from sowing which do not meet the standards on either way.

Methods of germination testing

A. Equipments (1) Counting Board
Germinations / Growth Chamber.

B. Sustrateem

1. Paper
2. Top of paper (T.P.)

3. Between Papers (BP)
4. Pleated Papers.

II Sand

Special requirement

1. Light Requirement – for seeds requiring light the tests should be illuminated for at least 8 hours per day at a high intensity of 750-1250 lux from white lamp.
2. Potessium Nitrate (KNO_3) - The germination substratum is saturated with 0.2% KNO_3 solution prepsred by dissolving 2 g KNO3 in one litre of H_2O.
3. Gibberellic acid (GA3) germinator substratum is dissolved in 0.01% to 0.1% GA3 solution. (100 mg – 1 g GA3).

Evaluation of germination test

1. **Normal seeding** – Seedlings which show the potential for continued development in to a normal plant. Normal seedlings most conform with one of the following categories.

A. Intact seedling – Seedlings with all normal essential structures, well developed, compete in proportion and healthy.

B. Seedlings with slight defects – Seedlings showing slight defects of their essential parts provided other wise have satisfactory growth and development.

C. Seedlings with secondary Infection

II Abnormal seedlings

Seedlings, which do not show the potential to develop in to a normal plant when grown under optimum conditions:

The following types of seedling are classified as abnormal.

- Damaged seedling seedlings with any of the essential structures missing on are irreparable that balanced development cannot take place.
- Deformed or unbalance seedlings - Seedlings with weak development or physiological disturbance or in deformed a not proportionate.
- Decayed seedlings - Seedlings with any of their essential structures so diseased/decayed as a result of primary infection that normal development is prevented.

III Hard seeds

Seeds of legumiansel and Malvaecae remain hard at the end of the prescribed test period because they have not absorbed moisture due to an impermeable seed coat.

IV Fresh un germinated seeds

Seeds other than hard seeds remain firm and apparently viable at the end of the test period. These imbibe moisture but do not germinate.

V Dead seeds

Seeds which at the end of the test period are neither hard nor fresh and do not produce any part of the seedlings.

Importance of seed vigor

Seed quality has traditionally been associated primarily with germination and purity. It is an important seed quality component, which needs attention. Seedlots having the same germination may differ in field emergence when sown at the same time in the same field and may also vary in germination when stored in the similar environment. Seed vigor refers to both the ability and strength of a seed to germinate successfully and establish successfully into a normal plant with optimum population under sub and supra-optimal conditions. The soil conditions during sowing time are usually not optimum and the common aim of the seed man or seed grower is to produce high quality seed and this on the seed production environment during seed development on the mother plant. The maximum seed quality is achieved at physiological maturity and later steadily dealings.

In order to maximize yield the seed procedure requires rapid establishment of seedling with optimum plant stand.

Performance of a crop is determined by the genotype x environment interaction. Reduction in seed viability may compensate the plant stand by adjusting or increasing the seed rate. Seed lots having low vigour when seeded in the field will have lower crop performance giving rise to

1. Low plant population
2. Lower seedling growth rate.
3. Plants developing from low vigour are prone to low performer and are unable to perform well under stress condition
4. There may be more numbers of non-productive plants.

What is seed vigor

ISTA defines seed vigor as an index of the extent of the physiological deterioration and/or mechanical integrity of a high germinating seed lot which governs its ability to perform in a wide range of environment. The reduction in the ability to the seed for its performance is known as physiological deterioration (ageing) and this begins before harvest and continence during harvest, processing transportation and storage, which is generally influenced by production environment. The end point of this deterioration is ultimately death seeds loose their vigor before losing their ability to germinate. Hence seed vigor can also be defined as

(1) The condition of the seed of active good health and robustness which on planting permits germination to proceed rapidly under a wide range of environmental conditions.

(2) The potential of rapid and uniform germination and fast seedling growth under general field conditions.

Manifestations of differences in seed vigor

The differences in seed vigor may be expressed throughout the plant life-cycle which are more apparent in some crops and are expressed in terms of germination or field emergence, speed of emergence, plant growth and yield, uniformity in plant stand, morphological abnormalities and seed longevity.

Germination

Seeds when tested in the laboratory for germination may differ in their performance in field condition due to sub optimal field conditions.

Besides germination the vigour differences are also expressed in the form of speed of germination and seedling growth rate. The longer the time taken for the seed germinate and to emerge there are greater chances of mortality due to physical stress or pathogen infection.

Plant growth and crop yield

During seedling emergence plant growth is nearly always exponential as there is nearly absences of compilation for growth resources. Hence a few laip delay in seedling emergence may result in difference in subsequent plant growth. The time of seedling emergence has its maximum effect on Spread is large,

When seedlings have high relative growth (warm moist conditions fast growing species)

When between plant competition is intense (at high plant densities) and when harvests are made not long after seedling establishment.

There is a possibility that the yield of plans grown from lower vigour seeds ay yield less than those grown having higher vigour.

Morphological abnormalities

A number of morphological abnormalities have been found to be associated with low vigour seeds viz.

- Loss of storage tissue
- Injury to meristematic tissues of plumule/ radide
- Damage to the vascular connections as transverse breaks in cotyledons and the presence of abnormal embryo.

Seed longevity

Differences in seed vigour also effect the seed longevity and storability even of those seeds having high germination.

Therefore seed vigour tests are important due to the following reasons.

1. The users can be alerted of the possible dealing in vigour and germination, between germination tests and sowing which can arise thropugh problems in production handling and storage.
2. These tests enable one to distingush seed lots, which are vulnerable to sub-optimal conditions and extremes of temerature and moisture.
3. It reduces the chances of crop failure at emergence, in other words high vigour seed lots can be identified for use in sub-optimal conditions.
4. These provide an opportunity to evaluate physiological seed treatments aimed at enchaining the performance of low vigour seed.

Seed vigour testing methods

Physical test

1. Seed size

1000/1000 sees drawn randomly and weighted in grams seed lot having higher seed weight are considered vigorous.

2. Seed density

Kerosene oil is placed in a graduated measuring cylinder initial level of the oil is noted. A weighed amount of seed is poured into the cylinders and the level of oil is noted difference between the two levels shows the density of the seed lot having higher density are considered to be vigorous.

3. Physical soundness

The seed lots having under sized, undeveloped, discolored shriveled and insect damaged seeds are considered to be weak.

Performance test under optimum conditions

1. **First count** – The normal seedlings counted at the first cont i.e. 4th – 5th day represents faster germination seeds maximum percentage of normal seedlings during first count indicates seed vigor.
2. **Speed of germination** – It indicates the vigor of the seed lot number of seeds germinated/day are counted and the cumulative index is made by the formula.

$$N^1/1 + N^2/2 \qquad tA / ny/y = N \text{ where}$$

N1 = Ny are the number of seeds germination day 1 to day.

1-y – as the number of days

High value of N indicates high seed vogour seed is considered as germinated when the radicals appear. Hence it should be counted daily and seeds observed as germinated should be removed.

Seedling growth rate

Paper towel is moistener with distill water twenty seeds are placed on it in a straight line and paper towel is folded and kept in a germinator at an angle of 75 and optimum temperature. Only 10 competitive seedlings are selected for observation and the rest are removed the length of each seedling is measured daily and seedling growth rate is determined as follows.

$SL_1 / F_1 (SL_1 - SL_2) / F_2 +$ .... $+ (SLn - SL(n\text{-}1)/F_n)$

Where SL_1 - Mean seedling length at first count

SL_2 - Mean seedling length at second coun

SL_1-SL_2 - Mean increase in length at second count

F_1 - Days of first count

F_n - Days to final count.

Seedling length

Length of 10 normal seedlings in cm in is observed during the final count grown id moist paper towel kept in a germinator at optimum temperature. The lot exhibiting maximum seedling length is considered vigorous.

Seedling dry weight

The weight of seedling excluding the cotyledons is taken on 10th day after drying them in an oven at 100oC for 24 hrs in g. The lot exhibiting maximum dry weight is considered as vigorous.

Strong and weak seedling

Seeds are placed on a moist paper towel at optimum temperature in an incubator. After 5 days the seedlings are categorized into strong and weak seedling. Seedlings are said to be weak when primary root, cotyledons or primary leaves are missing or short or missing primary leaf, spindly or poorly developed seedlings.

Vigor index by length

A combination of standard germination test with seedling length proveds evaluation for seed vigor (Abdul Baki and Anderson 1973)

Vigor Index = Germination x seedling length at final count.

Vigor index by weight is determined by the multiplication of seedling dry weight on the final count and germination percentage.

Seed metabolic efficiency (SEM)

It is the amount of dry seed weight respired for producing 1 g of dry root and shoot. Thus lighter the value of SME lower the seed efficiency as more amount of seed reserve would be used to for producing root and shoot. Amount of food material respired is (RESP) calculated.

RESP = SDW – (SHW + RTW + RSW)

Where

SDW = Seed dry weight before germination

SHW = shoot dry weight

RTW = Root dry weight

RSW = Seed dry weight after germination

Mibilization efficiency (MF)

Seeds lots having higher mobilization efficiency are considered as vigorous due to its capacity to supply food material to the growing seedlings. Weighed seeds are placed on towel paper and kept in germinator at an optimum temperature on the day of final count seedlings and cotyledons are dried separately at 100oC for 24 hrs. ME is calculate as follows.

$$ME = \frac{\text{Increase in dry weight of embryonic axis (dry weight of seedling)}}{\text{Decrease in dry weight of cotyledons}} \times 100$$

Stress tests

To date no one vigor test has been considered universally acceptable for all crops. However for some vigor tests research continues to advance their standardization.

Accelerated ageing test

The accelerated ageing test is the second most popular seed vigor test (Ferguson and Spear 19950 It provides valuable information on both seed storage and field emergence potential. It is applicable to a number of crops considerable efforts have been made to understand accelerated ageing test variables and standardize the test protocols (Tomes, TeKrony 1993: 19995) and incorporated into the ISTA Vigor Test handbook as a recommended vigor test for soybean. For sapid determination of seed vigor and storage potential of the seed lot the seed age is accelerated by weeks or days by subjection the seeds to high humidity and high temperature conditions. Seeds are spread on the wire mesh in single layer and place in a dissector previously filled with water a little below the wire mesh. The dissector is made airtight by applying grease to its lid and kept in an incubator at 40o C. Samples are drawn periodically to test the germination till it falls below 50%. The seed lots showing high germination for a longer period are considered to be vigorous.

Controlled deterioration

The controlled deterioration test has been recommended for vegetable crops and is considered n the ISTA Vigor test handbook. (Hampton and TeKrony 1995). This test is similar to accelerated ageing as it deteriorates low quality seeds at high temperatures and elevated seed moisture content. It differed from it by using seeds imbibed in water to a precise level allowing. The moisture to equilibrate by sealing the seeds in aluminum packets in refrigerator overnight and then submerging the seeds packets in a water bath a 45° C for 24 hrs (Powell 1995). The moisture range varies between 19-24%.

Brick gravel teat (Hiltner Test)

Seeds are place at a depth of 3 cm in tightly packed crushed brick granules of about 2-3 mm sizes. As the seedlings emerge the brick granules create a stress for the emerging seedling and impede the emergence of weak and mechanically injured seedlings.

Paper piercing test

Seeds are place on 1.25 cm of moist sand and is covered with a dry filter paper of 0.4 mm thickness 4-bulk, 90 g / mt m basic weight 0.39 kg/cm dry bursting strength, 1000-5000 m breaking length, 500 ml/minute filtering speed, 1500 mm wet bursting strength and 0.1% fiber composition, chemical wood pulp with high alpha percentage. This filter paper is again covered with 3 cm of moist sand. The tray is kept at 20-25o C up-to the day required for final count seedlings, which are able to penetrate the paper, are said to be vigorous lot.

Compact soil test

Seeds are planted in soil at optimum conditions in a tray. It is covered with a uniform and compact layer of the same soil. The seed lot having maximum emergence are considered to be vigorous.

Cold test

The cold test of maize is the most popular vigor test in USA (Ferguson and Spears 1995) is difficult to standardize among laboratories. This is mainly attributed to variability in soil moisture content where reduction in seed germination occurs above 34% soil moisture content (Nijenstein 1995) Small differences in temperatures during cold test also adversely affect the result.

Seeds are planted on 2 cm thick-leveled most soil. The soil is place on these seed. Enough cold water (10° C) is added to the soil to make it most then incubated at 10o C for 7 days. Later it is transferred at optimum required temperature for germination. Another set of the some lot should be kept simultaneously without any check. The lot showing minimum variation in germination when compared may be considered as vigorous.

Cool germination test

Seeds are placed on moist towel paper or sand and incubated at low temperature (10-15° C) up to the day of final count. The lot having maximum germination, seedling length, and dry weight are considered to be vigorous.

Biochemical test

1. Electrical conductivity test

Conductivity test is done to determine seed vigor does not work for all crops, which may be due to the presence of semi-permeable membrane that permits the inward entry of water but not the outward diffusion of certain electrolytes. In other cases lack of success has been attributed to genotypes, as high sugar sweet corn genotypes that possess thinner pericarps may be more susceptible to damage during harvest and drying (Waaes 1986). More emphasis has been given on what is leaked from the seed rather than relying on composite conductivity readings. The solutes leaked out from seeds when immersed in water are not available to the respiring and germinating seed, which result in poor emergence (Mathews and Whitbread 1968). A deteriorated seed lot leaches more water-soluble compounds the electrical conductance of a solution will be higher with higher concentrations of ions in the solution.

Equipments and chemicals

Analytical balance, beaker, de-ionized water, incubator, and electrical conductivity meter.

Procedure

50 seeds in three replications from each sees lots are weighed to two decimal place and placed in a beaker with 250 ml de-ionized water. These beakers are kept at 20°C after proper covering to reduce the evaporation and contamination. A beaker containing de-ionized water with no seed is kept each set is kept as control. After 24 hours water is poured through a coarse sieve into another beaker.

The conductivity of water in control is subtracted from the reading of soak water before. Calculation the conductance per gram of seed and suppressed as micro siemens /g of seeds (u/cm/g).

Interpretation

24 or less – Normal Sed; 25-29 suitable for early sowing, risk of poor performance under adverse condition; 30-43. Not suitable under adverse condition 44 and more – not suitable for sowing.

2. Tetra-zolium test

Equipment and chemical

The activity of dehydrogenses is expected to be more in vigorous seed (Lakon 1942, More 1973)

Procedure

Seeds are place on petri-plates lined with moist blotter paper at 25 + 1o C for 24 hrs. Embryonic axis is excised and kept in 1 ml of 15 (w/v) tetrazolium solution for 2 hrs at 30 + 1°C in dark. Excess solution is drained out and seeds are washed thoroughly with distilled water. The axis is soaked in 10ml of methyl allosove (methxy cellosol) for 4-6 hours with occasional stirring till the formazon is completely extracted i.e. axis turns colourless. Extract is decanted and its colout intensity is read at 480 n, in spectrophotometer.

3. Volatile aldehyde test

200 seeds are place on eight Whatman No. 1 filter paper of 7 cm diameter and moistened with 125 ml of distilled water and place in a 150 ml Erlenmeyer flask. A tube containing 10 ml of 0.2% (w/v) 3 methyl 2 benzothiozolium hydrogen hydrochloride monohydrate (MBTH) is placed in each flask. The flasks are sealed with rubber stoppers and incubated at 25 + 2°C for 24 and 48 hours respectively. 1.0 ml aliquot is pipeted into 2.5 ml freshly prepared 0.23% ferric chloride solution, 6.5 ml acetone, is added to each

tube after 5 minute to exhaust the reagent the enhance the color development. Absorption of the color developed is read on 720 spectrophotometer at 623 nm using fresh MBTH solution s blank. The values of absorbance are converted to mg formaldehyde equivalent using the standard curve for formaldehyde.

BIOCHEMICAL COMPOSITION OF MEDICINAL AND AROMATIC PLANTS

V.K. AGRAWAL
Professor and Head
Department of Plant Physiology, JNKVV, Jabalpur (M.P.)

Introduction

Plants have been one of the important source of medicines even since the dawn of human civilization. In spite of tremendous developments in the field of allopathy during 20th century, plants still remain one of the major sources of drugs in modern as well as traditional systems of medicine throughout the world. Approximately one-third of all pharmaceuticals are of plant origin, wherein fungi and bacteria are also included. Over 60% of all pharmaceuticals are plant based.

Medicinal plants have curative properties due to the presence of various complex chemical substances of different composition, which are found as plant secondary metabolites in one or more part of these plants. These plant metabolites according to composition are grouped as-alkaloids, glycosides, steroids or other groups of compounds which may have marked pharmaceutical actions as anticancerous, antimalarial, anihelmenthic or antidysentric, etc.. Many of the essential oils, dyes, lattices and even tannins and vegetable oils are also widely used as medicines.

Active constituents medicinal plants

Active ingredients are the main effective compounds of medicinal plants. Quantification of Active ingredients is the first step of stanndardiztion. Various chemical tests are available for quantification of active ingredients. Some of the main groups of active ingredients are described here.

Mucilage

Mucilage found in many plants is made up of polysaccharides (large sugar molecules) that soak up water, producing a sticky jelly-like mass. Mucilage lines the mucous membranes of the digestive tract, protecting against irritation, acidity and inflammation. This soothing and protective action appears to extend to other areas, including the mucous membranes of the throat, lungs, kidneys and urinary tubules. Slippery elm *(Ulmus rubra),* is a typical mucilaginous herb.

Phenols

This group of compounds includes salicylic acid-the natural forerunner of aspirin./Salicylic acid is found in many plants, for example wintergreen *(Gaultheria procumbens),* and white willow *(Salix alba).* Another phenol is thymol-a constituent of thyme *(Thymus vulgaris).* Phenols are antiseptic and reduce inflammation when taken internally yet they have an irritant effect when applied to the skin.

Tannins

Tannins are produced to a greater or lesser degree by all plants. The harsh, astringent taste of tannin-laden bark and leaves makes them unpalatable to insects and grazing animals. Tannins contract the tissues of the body and draw the tissues closer together to improve their resistance against infection. Oak bark *(Quercus* spp), and black catechu *(Acacia catechu)* are high in tannins.

Coumarins

Coumarins of different kinds are found in many plant species and have widely divergent actions. The coumarins in melilotus (*Melilotus officinalis*), thin the blood, while bergapten, found in celery *(Apium graveolens)*, is used as a sunscreen, and khellin, found in visnaga (*Ammi visnagal*), is a powerful smooth muscle relaxant.

Anthraquinones

Anthraquinones are the main active constituents in herbs such as senna *(Cassia angustifolia)*, and Chinese rhubarb *(Rheum palmatum)*, both of which are taken to relieve constipation. Anthraquinones have an irritant laxative effect on the large intestine, causing contractions of the intestinal walls and stimulating a bowel movement approximately 10 hours after being taken. They also make the stool more liquid, easing bower movements.

Flavonoids

Found in many plants, flavonoids have a wide range of actions. They are anti-inflammatory and are especially useful in maintaining healthy circulation. Rutin, a flavonoid found in plants including buckwheat and lemon, strengthens capillary walls.

Anthocyanins

These pigments, which give flowers and fruits a blue, purple or red hue, help to keep the blood vessel; healthy. Blackberry *(Rubus fruticosus)*, and grapes *(Vitis vinifera)* contain appreciable quantities anthocyanins.

Glucosilinates

Found exclusively in species of the mustard family, glucosilinates have an irritant effect on the skin, causing inflammation and blistering. Applied as poultices of painful or aching joints, they increase blood flow to the affected area, helping to remove the build-up of waste products la contributory factor in joint problems). Glucosilinates also help to reduce thyroid function. Both radish and mustard contain significant quantities of glucosilinates.

Volatile oils

Volatile oils-which are extracted from plants to produce essential oils-are some of the most important plant constituents of all. Tea tree *(Melaleuca altemifolia),* for example, is known to contain over 60 different volatile compounds within its volatile oil, many of them being strongly antiseptic. Some volatile oils contain sesquiterpenes, such as azulence, found in German chamomile. These constituents have an anti-inflammatory effect.

Saponins

There are two types of saponins-triterpenoid and steroidal saponins. The latter get their name from their similarity to the human body's own naturally occurring steroid hormones. Many plants containing steroidal saponins have a marked hormonal activity, liquorice *(Glycyrrhiza glabra),* being one of the best known. Triterpenoid saponins, for example those in cowslip root *!primula veris),* are often strong expectorants, and may also aid in the absorption of nutrients.

Cardiac glycosides

Found in various medicinal plants, most famously in common foxglove *(Digitalis purpurea),* yellow foxglove *(D. lutea),* and woolly foxglove *ID. lanata),* cardiac glycosides'such as digitoxin, digoxin, and gitoxin have a strong, direct action on the heart, helping to support its strength and rate of contraction when it is failing. Cardiac glycosides are also significantly diuretic. They help to transfer fluids from the tissues and circulatory system to the urinary tract, thereby lowering blood pressure.

Cyanogenic glycosides

Though these glycosides are based on cyanide, a very potent poison, they have a helpful sedative and relaxant effect on the heart and muscles in small doses. Wild cherry bark (Prunus *serotina),* and elder *(Sambucus nigra),* both contain cyanogenic glycosides, which contribute to both plants' ability to-suppress and soothe irritant dry coughs.

Vitamins

Some plants contain significant levels of vitamins. Watercress *(Nasturtium officinalis)*, for example, contains an appreciable quantity of vitamin E, and the hips of dog rose *(Rosa canina)*, have particularly high levels of vitamin C. Most other medicinal plants contain at least some vitamins. While the content may be small it contributes to overall daily intake. For other plants that are rich in vitamins.

Bitters

Bitters are a varied group of constituents linked only by their pronounced bitter taste. The bitterness itself stimulates secretions by the salivary glands and digestive organs. Such secretions can dramatically improve the appetite and strengthen the overall function of the digestive system. With the improved digestion system. With the improved digestion and absorption of nutrients that follow, the bodyis nourished and strengthened. Many herbs have bitter constituents, notably wormwood *(Artemisia absinthium)*, and chiretta *(Swertia chirata)*.

Alkaloids

A very mixed group, alkaloid mostly contains a nitrogen-bearing molecule (-NH21 that makes them particularly pharmacologically active. Some are well-known drugs and have a recognized medical use. Vincristine, for example, derived from Medagascar periwinkle *(Vinca rosea)*, is used to treat some types of cancer. Other alkaloids, such as atropine, found in deadly nightshade *(Atropa belladonna)*, have a direct effect on the body, reducing spasms, relieving pain and drying up bodily secretions.

Minerals

Some herbs are particularly rich in minerals. Horsetail *(Equisetum arvense)*, for example, has high levels of silica.

Dandelion *(Taraxacum officinalis)*, has large quantities of potassium, and unlike other diuretics, which flush this mineral out of the body, it helps to maintain high levels of potassium. These plants act as mineral supplements in their own right, while other herbs with a small concentration contribute to overall intake. For other plants with a high mineral content.

However, it should be stated in all fairness that our knowledge of the genetic and physiological makeup of most of the plant is poor and we know still less about the biosynthetic path ways leading to formation of active constituents for which they valued.

Out of nearly 3,00,000 species of higher plants available, only a small proportion have been investigated for medicinal properties, and a still smaller number yield well defined drugs. Thus, the knowledge of plant constituents gained so far is still meagre, considering the huge number species available in the world. Approximately, only 10% of the organic constituents of plants are reported to be known and the remaining 90% are yet to be explored.

Drugs are derived from trees, shrubs and herbs and even from primitive kinds of plants, which do not fall into the above categories. They are made from fruits (Senna, *Solanum viarum*, Datura etc.), flowers (*Butea monosperma, Bauhinia verigata*), leaves (Senna, Datura, Periwinkle, *Tylophora* etc.), stems (Liquorice, Ginger, Dioscorea, Costus, Garlic), roots (Rauvolfia, Periwinkle, Ginseng, etc.), seeds (Isabgol, Abrus, *Nux vomica*) and even bark (Cinchona).

The Indian Contribution

The Aryans of the Indus Valley wrote three treatises, viz. The ***Rigveda*** (2000 BC), ***Atharvaveda*** (2000-1000 BC) and ***Ayurveda*** (100-600 BC) which mention several medicinal plants and their uses. The ***Charaka Samhita*** , an encyclopaedia of Indian medicine, published at Varanasi between 1000 BC and 100 AD is a comprehensive record of medicinal plants and their uses. The indigenous systems of medicine practiced in India are based mainly on the use of plants ***Charaka Samhita*** (1000 BC – 100 AD) records the use of 2000 vegetable remedies. Ancient medicine was not solely based on empiricism and this is evident from the fact that some medicinal plants which were used in ancient times still have their place in modern therapy.

Status in India

The age-old Indian systems of medicine have been neglected mainly because of the rapid expansion of the allopathic system of medical treatment. Presently, the Indian system of medicine uses over 1100 medicinal plants and most of them are collected from the wild regularly, of which over five dozen species are said to be in great demand. There are around 10000 licensed pharmacies practicing the Indian system of medicine today, whose requirements have to be met. There are thousands of local vaidyas, herbal healers, monks, bone-setters and tribal doctors who practice this system of medicine. Many of them depend on retail outlets for their supply of raw materials.

There is a tremendous demand for raw material from all the four segments of the drug industry :

i. Plant drugs for Indian systems of medicines covering the Ayurvedic, Unani and Siddha systems,

ii. over – the – counter non-prescription items involving plants, extracts and Galanicals,

iii. Essential oils and

iv. Phytopharmaceuticals. Some of the medicinal plants growing in the wild are becoming extinct on account of destructive collection techniques, particularly when the whole plant is uprooted. Large scale use of adulterants and other substitutes reduce the potency and efficiency of the formulations.

Most of the Ayurvdic physicians regard *Centella asiatica, Merrimia emerginata* to be Brahmi while *Bacopa monneri* is the real Brahmi of Ayurveda. Similarly , *Chaeraphyllum villosum* sold in place of *Acntium heterophyllum* as' Atis'. *Polyalthia longifolia* and *Shorea robusta* in place of Soraca as Ashok

a) World Status

A vast majority of the world population today are finding themselves unable to afford the products of the western pharmaceutical industry, and they have to depend mainly upon the use of traditional medicines. This reality has been recognized, documented and compiled by the WHO in an inventory of medicinal plants numbering over 20000 species. WHO encourages, recommends and provides for the inclusion of herbal medicines in natural health care programmes. As they are easily available at cheaper price, time tested and considered safer than some of the modern synthetic drugs, there has been a resurgence of interest in the Western European countries.

Reasons for high demand of herbal medicines

The market for herbal medicines in the developing countries is growing at a faster rate than other pharmaceutical products. Some of the reasons are:

(i) The realization that allopathic drugs have harmful side effects.

(ii) Allopathic medicines are said to be ineffective against many chronic diseases like cancer. Moreover, many people suffering from diabetes, arthritis, respiratory diseases, skin ailments, gastric problems, jaundice, etc. are said to be turning more and more to Ayurveda and Unani for permanent cures.

(iii) Herbal medicines are comparatively less expensive.

(iv) The Western medical profession has begun to acknowledge the value of herbal medicines. This also explains the fact that many purely allopathic units have also adopted the use of formulations, which include natural herbal drugs.

The Development of Pharmacopoeia and Chemical Investigations

Today, chemical and pharmaceutical investigations have added a great deal of status to the use of medicinal plants by revealing the presence of the active principles and their actions on human and animal systems. Investigations in the field of pharmacognosy and pharmacology have supplied valuable information on medicinal plants with regard to their availability, botanical properties methods of cultivation, collection, storage, commerce and therapeutic uses. All these have contributed towards their acceptance in modern medicine and their inclusion in the pharmacopoeias of civilized nations.

Medicinal plants and Phytopharmaceuticals

The world pharmaceuticals Industry is unique, extremely competitive and highly research based. It is characterized by some very special and striking features like : high rate of obsolescence of drugs and technologies; requirements of high quality of products; Good manufacturing practices (GMP), Good Laboratory practices(GLP), environ mental and regulatory requirements etc. the pharmaceutical research on a established parameters is a complex, multidisciplinary and time consuming activity. The current estimate indicates that more than 12 years of research efforts from discovery to marketing with average cost of about 250 US $ are required for a drug molecule to be used as therapeutic agent.

During the past decades, bulk production of plant based medicine has become an important segment of Indian pharmaceutical industry. Some of the phytopharmaceuticals which are produced in India at present include : morphine, codeine, papaverine (*Papaver somniferum), Quinine, Quinidine, Cinchonine and Cinchonidine (Cinchona spp. Viz., C. calisava, C.ledgeriana, C. officinalis and C. succirubra*) : hyoscine and Hyoscyamine (*Hyoscyamus niger and H. muticus*); Colchicine (*Gloriosa superba, Colchicum luteum* and

Iphgenia stellata): Cephaeline and emetin (*Cephalis ipecacuanha*) : Sennoside A and B (*Cassia angustifolia* and *C. acutifolia*): Rescepine, Rescinnamine, Ajmaline and Ajamalicine (*Rauvolfia serpentina*) : Vinblastine and vincristine, ajamalcine, raubacine (*Catharanthus roses*): Guggal - lipid (*Cammiphora wightii*), Taxol (*Taxus baccata*), Artemisinin (*Artemisia annua*) etc.

In the present time, pharmaceutical and chemical sciences have greatly contributed in the enhancement of the utility of medicinal plants. Selected medicinal plants have been subjected to regorous chemical analysis and the bioactive compounds have been isolated and variously evaluated. As a result, new drugs have been discovered and new applications have been found for the compounds/ medicines that were already in use.

Over the years, there have been four principal routes by which plants have led to new therapeutic agents.

1. Chemical constituents such as digitoxin, strophanhin, morphine and atrophine isolated from the plants are directly used as therapeutic agents.
2. Some of these plant constituents are used as starting material for the synthesis of useful drugs. For example, adrenal cortex and other steroid hormones are generally synthesized from plant steroidal sapogenins.
3. Some natural products of medicinal plants serve as models for pharmacologically active compounds in the field of drug industry. Frequently, the side effects of a natural product often prevent its use in medicine and resolved only by preparation of synthetic derivative - for example - cocain which led to the development of modern local anesthetics; modification of colchicine and of podophyllotoxin to obtain anti tumor preparation.
4. There has been use of plant constituents as research chemicals particularly in the areas of molecular biology. One of the major criteria for selection of plants

for its constituents which are used as research chemicals for discovery of new lead molecules is the utilization of available knowledge in coded or uncoded form in traditional or indigenous systems of medicines, including folk or tribal medicines.

In this context, the development of new drugs, viz., Reserpine and related alkaloids for treatment of essential hypertension and in certain neuro-psychiatric disorders from ***Rauvalfia serpentina*** (Serpagandha); Guggulpid – for lowering of cholesterol from ***Commiphora wightii*** (Guggal); Forskoline (Coleonol) as antihypertensive agents from ***Coleus Forskohlii***; Bacosdes – as memory enhancer from ***Bacopa monnieri*** (Brahmi), are some of interesting examplesof how observation is recorded in ancient medical texts of Ayurveda or folk medicine, when investigations by modern scientific methods led to the successful development of new drugs and also provided a lead compound for further modifications. Some traditional medicinal plants on which significant research leads have been obtained are:

Traditional Medicinal Plants on which Significant Research Leads have been obtained and have yielded clinically useful drugs

S.N.	Species	Common names	Drugs	Action / clinical use
1.	*Acorus calamus*	Vacha	Asarone	Tranquilizer
2.	*Adhatoda vasica*	Vasa	Vasicine (Pegarine)	Oxytocic
3.	*Andrographis paniculata*	Kalmegh	Andrographolide Neo- Andrographolide	Bacillary dysentery
4.	*Asparagus racemosus*	Satawar	Sapogenin glycoside	Anti-oxytocic, anti - Ulcerogenic, burns and wound healing.
5.	*Azardirachto indica*	Neem	Azadirachtin oil	Insecticide, antifertility, antidiabetic agent
6.	*Bacopa monnieri*	Brahmi	Bacosides	Antianxiety agent Improve intellect, Adaptogenic
7.	*Carica papaya*	Papita	Chymopapain	Proteolytic , mucolytic
8.	*Centella asiatica*	Brahmi	Asiaticoside	Vulenerary wound Healing.
9.	*Cissampelos pareira*	Patha	Cissampeline	Skeletal muscle relaxant
10.	*Coscinium fenestratum*	Daruhaldi	Alkaloids (Berberine)	Cholera and Gastroentritis
11.	*Curcuma longa*	Haldi	Curcumine	Choleretic
12.	*Gloriosa superba*	Kalihari	Colchicine, Thio-	Odema
13.	*Gossypium spp.*	Kapas , cotton	Gossypol, colchicoside	Male contraceptive
14.	*Mappia foetida*	Ghaneru	Camptothecein	Anticancer

S.N.	Species	Common names	Drugs	Action / clinical use
14.	*Mappia foetida*	Ghaneru	Camptothecein	Anticancer
15.	*Mucuna pruriens*	Kewanchh	L-dopa	Antiparkinsonism
16.	*Picrorhiza kurroa*	Kutki	Glycosides	Antihepatotoxic
17.	*Phyllanthus amarus*	Bhumyamalki	Lignans	Antihepatotoxic (Hepatitis B-Virus)
18.	*Silibum marianum*		Silmarin	Antihepatotoxic
19.	*Taxus baccata*	Talishpatra	Taxol	Anticancer
20.	*Tephrasia purpurea*	Sarpunkha	Flavonoids alkaloids	Viral hepatitis
21.	*Urginia indica*	Indian squill	ScillarinA and B	Cardiotonic
22.	*Veleriana wallichii*	Tagar	Valeopotriates	Sedative

Besides this the plant species viz., *Canarium euphyllum, C.manii* (collected from Andaman forests), *Paconia obovato* (collected from western ghats) are found to be very active against *rhematoid arthritis* and *inlammation.*

POST HARVEST TECHNOLOGY OF MEDICINAL AND AROMATIC PLANTS

A. UPADHYAY
Associate Professor
Deptt. of Plant Physiology,JNKVV, Jabalpur (M.P.)

In recent year there has been a tremendous growth of interest in plant based drugs, Pharmaceuticals, perfumery products, cosmetic, and aroma compounds used in food flavors and fragrances and natural colours in the world.

Furthermore, active constituents are often more highly concentrated in wild plants, as the herb is generally likely to be growing in its preferred environment.

Collection

Medicinal and aromatic plants are so far collected following the traditional methods without adhering to the pharmacopoeia standards. Collection of plants in wild continues to be unscientific, labour-oriented and undependable. Many of the crude drugs are poisonous and dangers of collection by untrained collectors are obvious, because even experts have difficulty in correct identification of some species unless they are in flower. Thus, no control is kept on the quality of the material.

The collection of underground parts like roots, rhizomes, tubers etc. is done by digging out the entire plant by means of kutki (*Picrorrhiza kurroo*), as may be available from July-August to March-April. In certain cases, the collectors have been busy collecting these herbs during summers *i.e.,* the growing season, without caring for its survival and regeneration. By doing so, on one side, the collection of immature and undeveloped parts/plants results into a product of inferior quality lacking proper active ingredients and on the other side, the life cycle of the plant is broken.

In case of large flowers, it is picked just after they have opened, usually during the spring or summer. Sometimes only specific parts of the flower are used, such as the petals of marigold (*Calendula officinalis*), while other flowers are used as a whole. Full bloom flowers or blossoms are not suitable for drug market. The fruits and berries are harvested in early autumn when rope but still firm. If allowed to become over-ripe, they may not dry properly. They can be picked individually or in branches.

Underground parts are storage organs for the plant and accumulate active principles during summer months. Roots and rhizomes of perennials are generally gathered after two to three years of growth. The roots of annuals are generally not collected. This applies to majority the drug plants collected from wild. However, cultivation, the roots or rhizomes are harvested as per the requirements and presence of active constituents *e.g.* turmeric, dioscorea, ginger, etc.

All the underground portions are collected, cleaned and dried in shade or ovens at 30 ^{0}C, Proper care should be taken to dry the materials *e.g.* gloriosa, squill, garlic, saffron, colchicum etc. Sap is collected in the spring as it rises, Collect milky juices or latex from plants such as dandelion (*Taraxacum officinale*), by squeezing the stems over a bowl. Wear gloves as latex or sap can be corrosive. The gel from Indian aloe (*Aloe barbedensis*) is scarped out after slicing the teat lengthwise and peeling backs the edges.

Harvesting

Harvesting is an important operation in cultivation technology, as it reflects upon economic aspect of the crude drugs. Harvesting can be done efficiently in every aspect by the skilled workers. It is, however, a laborious job and may not be economical. In certain cases, it cannot be replaced by any mechanical means, *e.g.* digitalis (*Digitalis Purpurea*), vinca (*Catharanthus roseus*) and senna (*Cassia angustifolia*) leaves. The under ground drugs like rhizome, roots, tuber, bulbs, corms etc. are harvested by mechanical devices, such as digger or lifters. Drugs which constitute al aerial parts are harvested by binders for economic reasons. Many a times, a special device known as seed stripper harvests flowers, seeds and small fruits. The technique of beating plants with bamboos is used in case of cloves. The seaweeds producing agar. Peppermint and spearmint are harvested by normal method with mowers, whereas fennel, corinder and caraway are uprooted and dried. After drying, either they are thrashed or beaten and the fruits are separated by winnowing. Sometimes, reaping machines are also used for their harvesting.

Drying

Drying consists of removal of sufficient moisture content of crude drugs, so as to improve its quality and make it resistant to the growth of microorganism. Drying also facilitate and pulverizing or grading of a crude drugs. In certain drugs, some special methods are required to be followed to attain specific standards. *e.g.* fermentation in case of Dalchini (*Cinnamomum zeylanicum*) bark and Indian gentian (*Picrorrhiza kurroo*). The slicing and cutting into smaller pieces is done to enhance drying, as in case of liquorice (*Glycyrrhiza glabra*), squill and calumba, the flowers are dried in shade so as to retain colour and volatile content. Depending upon the type of chemical constitutes, a method of drying can be used for a crude drug. Drying can be of two types: (i) Natural (Sun drying), (ii) Artificial.

Natural Drying (Sun-drying)

In case of natural drying, it may be either direct sun drying or in the shed. If the natural colour of the drug (e.g. Digitalis, clove, senna) and the volatile constituents of the drugs (peppermint) are to be retained, drying in shed is preferred. If the contents of the drugs are quite stable to the temperature and sunlight, the drugs can be dried directly in sunshine (e.g. gum, acacia, seeds and fruits).

Artificial Drying

Drying by artificial means includes drying the drugs in (a) an oven; i.e. tray-dryers, (b) vacuum dryer and (c) spary dryers.

Tray Dryers

The drugs, which do not contain volatile oils and are quite stable to heat or which need deactivation of enzymes are dried in tray dryers. In this process, hot air of the desired temperature is circulated through the dryers and this facilities the removal of water contents of the drugs (e.g. belladonna roots, cinchona bark, tea leaves and gum are dried by this method).

Vacuum Dryers

The drugs which are sensitive to higher temperature are dried by this process e.g. tannic acid and digitalis leaves.

Spray Dryer

Drugs which are Highly sensitive to atmospheric conditions and also to temperature of vacuum-drying are dried by spray dryer method. *e.g.* papaya latex, pectin, tannins, etc.

Garbling (Dressing)

Preparation of crude for market after drying is garbling. Sand, dirt and foreign organic parts of the same plant, not constituting drug are required to be removed. This foreign organic matter (extraneous matter) is removed by several ways. Excessive stems in case of lobelia and stramonium (Datura spp.) need to be removed, while stalks, in case of cloves are to be deleted. Drugs constituting rhizomes need to be separated carefully from the roots and rootlets and also stem base. Pieces of iron must be removed with the magnet in case of castor seeds before crushing and by shifting in case of vinca and senna leaves. Pieces of bark should be removed by peeling as in gum acacia.

Packing

The morphological and chemical nature of drug, its ultimate use and effects of climate conditions during transportation and storage should be taken into consideration while packing the drugs. Aloe is packed in goatskin. Asafetida in stored in well closed containers to prevent loss of volatile oil. Cod liver oil, being sensitive to sunlight, should be stored in such containers, whereas the leaf of the drugs like senna, vinca and other are pressed and baled. The drugs are very sensitive to moisture and also costly at the same time need special attention. e.g. digitalis, ergot and squill. The squill becomes flexible, ergot becomes susceptible to the microbial growth, while digitalis losses its potency due to decomposition of glycosides. Cinnamon bark, which is available in the form of quills, is packed one side the other quill, so as to facilitate transport and to prevent volatilization of oil from the drug. The crude drug like roots, seeds and others do not need special attention and are packed in gunny bags, while in some cases bags are coated with polythene internally.

Storage

Preservation of crude drugs needs sound knowledge of their physical and chemical properties. A good quality of the drugs can be maintained. The should be stored in the premises which are waterproof, fireproof and rodent proof. A number of drugs absorb moisture during their storage and become susceptible to the microbial growth. The moisture, not only increases the bulk of the drug, but also causes impairment in the quality of crude drug. The excessive moisture facilitate enzymatic reactions resulting in decomposition of active constituents. *e.g.* ergot, cod liver oil and digitalis.

Apart from protection against adverse physical and chemical changes, the preservation against insect or mould attacks is also important. Different types of insects, nematodes, worms, mould and mites infest the crude drugs during storage more important pest found in drugs can be prevented by drying the drug thoroughly before storage and also by giving treatment of fumigants. The common fumigants used for storage of crude drugs are methyl bromide, carbon disulphide and hydro cyanic acid.

Temperature is also very important factor in preservation of the drugs, as it accelerates several chemical reactions leading to decomposition of the constituents. Hence, most of the drugs need to be preserved at a very low temperature the costly phyto-pharmaceuticals are required to be preserved at refrigerated temperature in well-closed containers. Small quantities of crude drugs could be readily stored in air-tight, moisture proof and light proof containers such as tins, cans, covered metal tins, or amber glass containers. Wooden boxes and paper bags should not be used for storage of crude drugs.

Value-Addition or Processing

In order to make use of its latent medicinal qualities, the active principles contained in every medicinal plant consist of a number of compounds, which individually or in groups, can have a specific action on the organs or our bodies. Different methods have, therefore, been devised to extract these substances, either singly or collectively, depending on the result required. Generally, there are four basic, simple preparations *viz.* decoction, maceration, infusion and juice extraction.

Decoction

Decoction is the method normally used for those medicinal herbs whose active principles are difficult to extract because they are contained in woody parts of the plant. Sometimes, the extraction be decoction involves boiling the whole plant or a part of it in water for a given time and allowing it to macerate for a further period before filtering. This method is reserved for certain plants such as those with a high mucilaginous content.

Extraction

Extraction is much more suitable method for obtaining the active principles when the parts of the plant being used are soft and fragile, such as leaves bud or flowers. It is preferable, except in certain cases, to use herbs that have been slightly dried, as the reduction of water content will concentrate the principal constituents.

Maceration

This method is used for medicinal plants whose active principles are soluble in cold water for several hours during, which time all the principles that do not need heat to release them (*i.e.* that are not thermo labile) will be released into the solution.

Freeze-Drying

Many plants may be juiced, frozen as ice cubes and thawed as required.

Preservation

Medicinal plants can be preserved in a number of ways, the most common and simple, being air or oven drying. A warm, dry place, for example an airing cupboard, is ideal Use plain paper for drying herbs, never printed newspaper. Dried herbs can be stored in a dark glass jar or a brown paper bag.

Active constituents

The active ingredients are the main effective compounds of medicinal plants. The quantification of active in gradients in the first step of standardization of crude drugs. A number of chemical test are available for quantification of active ingredients. Some of the main chemical groups (organic compounds) of active ingredients are:

(1) Saponins

They are often strong expectorants, and stay also aid in the absorption of nutrients.

(2) Cardiac Glycosides

They are diuretic. They help to transfer fluids from the tissues and circulatory system to the urinary tract, thereby lowering blood pressure.

(3) Flavonoids

It possesses anti-inflammatory properties and are especially useful in maintaining healthy circulation.

(4) Mucilage

It is made up of polysaccharides (large sugar molecules) that soak up water, producing a sticky jelly-like mass. Mucilage lines the mucous membranes of the digestive tract, protecting against irritation, acidity and inflammation.

Phenols

This group of compounds includes salicylic acid. Phenols are antiseptic and reduce inflammation.

Tannin

These are produced to a greater or lesser degree by the plants. They drawn the tissue closer together and improve their resistance to infection.

Coumarins

Coumarins of different kinds are found in many plant species and have widely divergent actions. It is a powerful smooth muscle relaxant.

Anthraquinones

Anthraquinones have an irritant laxative effect on the large intestine.

Anthocyanins

Help to keep the blood vessels healthy.

Glucosilinates

These compound are found exclusively some medicinal plants belong to mustard family. Glucosilinates have an irritant effect on the skin, causing inflammation and blistering. Glucosilinates also help to reduce thyroid function.

Volatile Oils

It is extracted from plants to produce essential oils. Some volatile oils contain sequiterpenes,. These constituents have an anti-inflammatory effect. And many of them being strongly antiseptic.

Cyanogenic Glycosides

Though these glycosides are based on cyanide, a very potent poison. They have a helpful sedative and relaxant effect.

Vitamins

Some plants contain significant levels of vitamins. Most other medicinal plants contain at least some vitamins.

Bitters

Bitterness itself stimulates secretions by the salivary glands and digestive organs. Such secretions can dramatically improve the appetite and strengthen the overall function of the digestive system.

Alkaloids

An alkaloid is a mixed group, containing nitrogen molecule (NH2) that makes them particularly pharmacologically active. Some of them are well known drugs and have a recognized medical use.

Minerals

A number of medicinal plants are rich in minerals, which helps in maintaining high levels of potassium.

Post Harvest Technology of some important medicinal and Aromatic plants

S. No.	Medicinal Plant	Plant part	Post harvest Technology
1.	Mulaithi	Root	After harvesting the roots, they are washed cut in to pieces of 15-20 cms, then dried at Temp. of 30-40 ^{0} C retaining 10% moisture.
2.	Kalihari	seed	Fruits are plucked when they begin to turn into Pale green colour and skin begins to shrink and dried. Then seeds are obtained from them carefully and stored.
3.	Safed Musli	Roots	Roots are dug at appropriate time. After sorting out the long and short roots, they are washed, skin is removed and dried either in sun light or solar drier keeping on white muslin cloth or on polythene sheet to prevent the colour from turning to off - white or Grey.
4.	Buch	Rhizome	Harvesting of rhizomes is done at the stage of senescence (yellowing of leaves). The bulbs are (Rhizomes) are cleaned without using water, cut into pieces and are dried in shade so that the oil content is not reduced.
5.	Sanay	Leaves	First harvest is at the stage of 90-100 DAS (Days After Sowing). After that second harvest is done50-60 days. Twigs are collected after plucking and left for drying on a Tarpaulin. Leaves are packed in gunny bags and stored.
6.	Isabgol	Seed coat	Harvesting is done at senescence (Ear heads turns to brown colour) stage in, morning hours. Then the crop is left for drying Sprinkling water at intervals. Seeds are carefully separated from beans, dried for 4-5 days and stored in gunny bags.

7.	Satawar	Root and Rhizome	Roots are dug after 3-4 years after planting. Then roots are graded lengthwise, washed sun dried and stored. Rhizomes are dried for 15-20 days because they contain 90% water in them.
8.	Giloe	Stem	Harvesting time is April-May. After harvesting, The stem is cut into horizontal and vertical pieces, dried in intense sunlight and packed in gunny bags and stored.
9.	Muskdana	Seeds	Crop harvesting time is 4-5 months after sowing. Ripened fruits (yellow in colour) are collected, dried and seeds are separated and stored for extraction of oil.
10.	Harra	Fruit	Seeds are collected from dry fruits and powdered for usage.
11.	Aonla	Fruit	Fruits are picked when their growth becomes standstill, colour turns to yellowish green or brown patches appear on the outer cover.
12.	Guggal	Branches	Gum is derived from the stem by putting a small incision on the stem. Gum can also be derived by cutting the branches and drying them.
13.	Arjun	Bark	After ten years of planting , bark can be obtained from the tree for use.
14.	Lemon grass	Grass	Leaves , Stem and Flower First harvesting can be done at 3-4 months after Sowing for green grass, which is used for extraction of oil. Final harvest of crop is done by leaving about 6" stubble from ground level.
15.	Java grass	Grass	Flower Harvesting is done in different stages i.e. from Flowering stage to seed maturity stage. I/3rd length of grass is cut

			from the ground level along with flowers leaving 1/4th part of stem on the ground. The grass is dried in shaded place, cut into pieces and put into distillation process for extraction of oil by vacuum evaporation.
16.	Serpagandha	Bark and Root	Root harvesting is done after 19-24 months after sowing. Care should be taken that the outer covering of roots is not destroyed while digging or washing of roots. Then the roots are cut into 10-15 cm pieces and dried.
17.	Ashwagandha	Roots	The crop is ready for harvesting when there is a plant population of about 8-10 lakh plants in one acre area, with yellowing of leaves and ripened red fruits. The plants are uprooted, washed , cut into pieces and dried in sunlight. Seeds obtained from dried fruits are stored .
18.	Satawar	Roots	Harvesting is done after 18 months of sowing , at the stage of senescence. At the same time disc is get readied for the next crop. Roots are washed, graded lengthwise. The outer cover of the taproots is removed carefully by putting an incision and dried in sunlight..
19.	Nagarmotha	Root	Harvesting of roots is done when the crop is 17-18 months old and the colour of leaves began to turn yellowish. After washing the roots, they are cut into thin wafers and dried in sunlight. After that, these wafers are put into extraction process by vacuum evaporation method.

IMPORTANCE OF QUALITY CONTROL OF MEDICINAL PLANTS

ANUBHA UPADHYAY
Associate Professor
Deptt. of Plant Physiology, JNKVV, Jabalpur.

Plant materals are used throughout developed and developing contries as home remedies, over the counter drug products and raw materials for the pharmaceutical injustry. With ever-growing commercialization of herbal drugs, there is an urgent need to develop some kind of standards and quality control parameters for the raw mateials as well as the medicines comprising single or multiple plant materials.

There are mainly three aspects for ensuring, quality control and standaridizationof plant drugs:

- Botanical identity & pharmacognostical examination of raw materals.
- Protecton of therapeutic potentials of the raw materials.
- Biological assay.

Botanical Identity and Pharmacognostical Examination of Raw materials

Before collection of a medicinal plant, it should be correctly identdied and one should know which part is required to be collected. For whole herb it is easy to identify by following taxonimical characters with the help of floras and matching them with authentic specimens preserved in various hervaria. But the isolated plant parts, such as roots, rhizome, tuberm bulbs, leaves and seed procured from the market are to be identified through pharmacognostic standards. It is, therefore, necessary to maintain such standards for each hebal drug used by the industry.

Protection Therapeutic Potential of the Raw materials

A medicinal plant from field to the drug preparation stage has to face various riddles particularly under storage. The collected plant materials are stored under unhygienic condition, which make these prone to microbial infestation which in turn deteriorate active principles of the crude sample.

Collection and Drying

Freshly collected chopped raw material has high moisture contend (55-90%); there fore, it has to be subjected to drying. Drying prevents moulding, the enzymatic action, and microorganism infestation and chemical changes. It fixes the constituents and facilitates grinding and milling as well as converting the raw drug into a more convenient form for industrial handling.

Storage and Preservation

Proper storage and presevation are important factors in maintaining a high degree of quality of raw materials. If these are not properly stored, they may absorb moisture which not only causes increase in weight of raw material, but also reduces the percentage of active principles. It also favours enzymatic action and hastens mecrogbial destructionof the material.

(a) Microbial Association with Crude Drug Samples Under Storage

The plants samples collected from forests are stored in traditional warehouse where they are usually packed in gunny bags or spread as such on ground and have to face fluctuation, environment and divers range of microbabes.

(b) Biodeterioration of Active Ingredients

Fungi are capable of infecting a wide variety of substrates due to their uniqiutous nature. During storage, the fungal organisms thrive on drug plants by utilizing various components including the active ingredients.

Biological Assay

The advantages of these methods are to be utilized for standardization of the final products, i.e. themedicines in respect of the following aspects:

To establish bioavailability parameters, chemical and biological.

- Toxicity studies, acute and chronic.
- Dose determination
- Clinical trials.

The future of the medicinal plants industry in Indian holds great promise since modern infrastructure and expertise are available. It is expected that in the coming years various methodologies would be adopted to bring revolutionary progress, which include genetic engineering, better crop mangement and improved processing technologies so that Inia could compete in the international market in a big way.

The credibility of the traditional systems of medicine, which have stood the test of time, will certainly go up the enhance their efficacy, feputation as well as acceptability, if quality control parameters are developed and strictly followed.

ECONOMICS EVALUATION OF MEDICINAL PLANTS

A .K. SARAWGI
Professor
Deptt. of Economics, JNKVV (M.P.)

Importance and Practical utility

In recent time due to the increasing realization of health hazards and toxicity caused by synthetic drugs and antibiotics, there has been a renewal of interest in the use of plant based drugs throughout the worl. The international trade on herbs and herbal drugs is growing substantially due to this awareness. India has been considered as treasure house of valuable medicinal and aromatic plants. The group of Medicinal and Aromatic Plants comprises a large number of plants species, which provide raw material for Pharmaceutical, Phytochemical, Food flavoring and Cosmetic industries. About 9500 plants species have been identified to have curative properties, aromatic or flavour. These plants are being used in Indian system of Medicine like Ayurveda, Unani, Siddha, Tribal Medicine and also in Homeopathy .The pharmaceutical and aroma industries use a large number of plants and new items are continuously being added to the list of commercial sources of industrial raw materials every year. Majority of the commercial supply of medicinal plants derived form the wild. Due to increasing awareness of the health hazards, the demand for herbal drugs are increasing day by day. India is bestowed with a unique wealth of medicinal plant resources. The foreign exchange also being earned by India from the export of medicinal and aromatic plants which is estimated to be over US $ 3000 million per annum. India at present export herbal material and medicines to the tune of Rs. 446.3 crores and according to the recent estimates, it can be raised 3000 crores by 2005. (Source: Planning commission, Govt. of India, march 2000).

India is estimated to have a botanical wealth of over 2000 medicinal and aromatic plants species. Rapid increase in population, expansion of area under food and commercial crops and deforestation led to considerable depletion of herbal wealth and even disappearance of certain valuable plant species particularly from their natural habitats. Unless effective measures are taken, major valuable species are likely to be lost forever in the near future .At present the demand for the these plant products shows a manifold increase in the domestic markets and substantial quantities are also exported in the crude form. Therefore cultivation and conservation of genuine plant species is very important. Although India is blessed with a variety of soil and agro-climatic situations that support a large variety of plants. So only alternative is to undertake large-scale cultivation and conservation of biodiversity of medicinal and aromatic plants than the extinction of valuable plant species of this group can be prevented. Cultivation of medicinal and aromatic plants is very profitable business but it needs proper planning as per the demand of raw material by the pharmaceutical and other allied industries at national and international level.

Due to huge demand of medicinal plant product and profitable enterprise, it must be considered in the cropping pattern as a cash crop by the grower / farmers. In this way it would generate income and employment opportunities, especially in the potential area of medicinal plant in Madhya Pradesh.

Considering the important of medicinal plants, it is necessary to know, how to estimate or evaluated cost of cultivation of medicinal crops. Now considered the most important medicinal crop i.e. Safed musli for this purpose. Before go into details about different cost concept which are used for estimation of cost of cultivation. First of all we must know important and practical utility of safed musli crop.

Safed musli

Safed musli is one of the important plant, which is found in nature. It is used in all type of medicines like Ayurvedic, Allopathic and Unani medicine.

Safed musli is richest source of

- Carbohydrates - 42%
- Proteins - 8 to 9%
- Saponins - 2 to 17%
- Alkaloids - 25%
- Vitamines - A, B, D, K. and E
- Glucocydes - Available
- Amino Acid - Available
- Staryords - Available
- Mineral matter - 7 to 15%

Uses of Safed musli

- Aphrodisiac Agent and Vitalizer
- As A General Sex Tonic
- Remedy for Diabetes
- As curve for Arthritis
- As A curative for Natal and Post Natal Problem
- For Rheumatism And Joint Pains

Safed musli is an important ingredient of more than a hundred Ayurvedic, Allopathic and Unani medicinal preparations. As a (protein rich) food supplement, there is a tremendously growing national and international demand for Safed Musli in the Middle East, Europe, Japan and the U.S.A. In world its availability is about 5000 tonnes while its estimated demand is 35000 tonnes per year.

It is found naturally in forest of our country but due to continuous distraction of the forest, safed musli is going to be indegard from the forest. There is a great need of cultivation of this important medicinal plant. Safed Musli grown successfully in many combinations and in varying climate, soil and irrigation conditions and it has grown in all the conditions even where we provide no care of weeding, manuring , irrigation etc. But in all the above cases, as far as yield was concerned, the results were different.

At present it is only possible to bridge the gap of demand and supply by scientific cultivation of Safed Musli. It is highly profitable crop then the other cereals, pulses and oil seed crops. About Rs. 2 to3 lakh can be obtained with one acre of Safed Musli. Safed musli is a cash crop with low risk and assured tax-free high return in a period as short as less than one year.

For economic evaluation (cost and return analysis) of Safed Musli crop, the following scientific cost concept may be used.

Cost concept

Cost $A_{1 == All}$ actual expenses in case and kind incurred in production by owner operator. Cost $A_{2 ==}$ Cost A_1 + rent paid for leased in land.

Cost B_1 = Cost A_2 + interest on value of owned fixed capital assets excluding land.

Cost B_2 = Cost B_1 + rental value of owned land.

Cost C_1 = Cost B_1 + imputed value of family labour.

Cost C_2 = Cost B_2 + imputed value of family labour.

Cost C_3 = Cost C_2 + 10% of cost C_2 as managerial cost.

Cost A_1 include value of hired labour, value of hired bullock labour, value of own bullock labour ,hired machinery charge, value of owned machine labour ,value of seed (both farm produced and purchased), value of insecticide and pesticide, value of manure (owned and purchase), value of fertilizer, depreciation of implements and farm buildings, irrigation charges , land revenue, cesses and other taxes, interest on working capital and miscellaneous expenses etc.

Profitability Aspect

1. Net farm Income (NFI) = Gross income – Cost C_3
2. Farm Business Income (FBI) = Gross income – Cost A_1
3. Family Labour Income (FLI) = Gross income – Cost B_2

Cost and Return Analysis of Safed Musli

i. Improved planting material.

ii. Wild planting material.

Cost of cultivation of Safed Musli for improved planting material

(A) **Expenses / Acre**

S.No.	Particular	Amount (Rs.)
1	Field preparation, leveling and preparation of beds	2000
2	Farm yard manure (10 Trolley) @Rs. 600 /Trolley	6000
3	Fertilizer, Spray and other soil tonics	2000
4	Green manure and mulching	3000
5	Planting material (4 quintal per kg. @ Rs. 400/Kg.	1,60,0000
6	Intercultural (four times)	2000
7	Sowing 10 labour @50/day/ labour	500
8	Entrooting and washing of tubers 20 labour	1000
9	Irrigation (2 Irrigation)	800
10	Miscellaneous expanses (Int. on working capital depreciation)	5000
11	Peeling and drying @ 100 / Kg. for about 400kg. Safed musli	40,000
	Total expenses	**2,22,300**

(B) Income (Per acre)

S.No.	Particular	Amount (Rs.)
1	Income of 4 Quintal dry Safed Musli @ Rs. 1000/kg.	4,00,000
2	Income in the form of planting material 4 qul. @Rs.300/kg.	1,20,000
	Total income (G.I.)	**5,20,000**

(C) Net Profit (Per acre)

Net Profit = Total income – Total expenses

5,20,000 - 2,22,300

Net Profit (Per acre)= 2,97,700

Cost of cultivation of Safad Musli for wild planting material

S.No.	Particular	Amount (Rs.)
1	Field preparation, leveling and preparation of beds	2000
2	Farm yard manure (10 Trolley) @Rs. 600 /Trolley	6000
3	Fertilizer, Spray and other soil tonics	2000
4	Green manure and mulching	3000
5	Planting material 80,000 Fingers/acre @Rs.150per kg. (4 quintal)	60000
6	Intercultural (four times)	2000
7	Sowing 10 labour @50/day/ labour	500
8	Entrooting and washing of tubers	1000
9	Irrigation	500
10	Miscellaneous expanses	5000
11	Peeling and drying @ 100 /Kg. For about 300kg. Safed musli	30,000
	Total expenses	**1,12,000**

(B) Income (Per acre)

S.No.	Particular	Amount (Rs.)
1	Income of 3 quintal dry Safed Musli @ Rs. 500/kg	1,50,000
2	Income in the form of 3 quintal planting material @Rs.150/kg.	45,000
	Total income	**1,95,000**

(C) Net Profit (Per acre)

Net Profit = Total income – Total expenses

1,95,000 - 1,12,000

Net Profit (Per acre)= 83,000

Economics of Planting Material of Safed Musli

Cultivation	Cost of Cultivation per Acre (Rs.)	Gross return per Acre (Rs.)	Net Return per Acre (Rs.)	Cost of Production per kg.	Return per rupee invested (RPRI)
Imputed planting material (IPM)	2,22,300	5,20,000	2,97,700	111.15	2.34
Wild planting material (WPM)	1,12,000	1,95,000	83,000	74.67	1.74
Yield IPM	20 Q/Acre				
WPM	15 Q/Acre				

Cost and Return Analysis of Safed Musli

(As per scientific cost concept)

For Imputed Planting Material

Cost A_1

S.No.	Particular	Amount (Rs.)
1	Field preparation (3 ploughing) by Tractior and leveling	1000
2	Farm yard manure (10 Trolley) @ Rs. 600 /Trolley	6000
3	Green manure and mulching	3000
4	Preparation of row and beds @Rs.50/day No. 20	1000

S.No.	Particular	Amount (Rs.)
5	Fertilizer, other soil tionic	2000
6	Planting material (4 quintal per kg. @ Rs. 400/Kg.	1,60,0000
7	Intercultural operation (four times)	2000
8	Sowing 10 labour @50/day/ labour	500
9	Irrigation (2 Irrigation)	800
10	Entrooting and washing of tubers 20 labour @Rs. 50/day	1000
11	Peeling and drying @ 100 /Kg. for about 400kg. Safed musli	40,000
12	Miscellaneous expenses (Land revenue, taxes, value of depreciation, Interest on working capital and other expanses)	5000
	Cost A_1	**2,22,300**

Cost A_2 = 2,22,300 + Nil

Cost B_1 =2,22,300 + 2000 = 2,24,300

Cost B_2= 2,24,300 + 3000 =2,27,300

Cost C_1 = 2,24,300 + Nil

Cost C_2 = 2,27,300+ Nil

Cost C_3 = 2,27,300 +22,730 =2,50,030

i. Net Farm Income (NFI) 5,20,000 – 2,50,030= 2,69,970

ii. Family labour Income (FLI) 5,20,000 - 2,27,300 = 2,92,700

iii. Farm Business Income (FBI) 5,20,000 – 2,22,300 = 2,97,700

Benefits of Safed Musli

a. Short duration and high value crop. Within 6-8 months it gives income of Rs.2-3 lakh per acre.

b. There is no need of safed musli by peeling and drying it can be directly sold in the market.

c. Training regarding cultivation of safed musli is available of several places.

d. It provides self-employment to the unemployment youth.

e. There is no effect of change in the weather on crop.

Problem of Safed Musli in cultivation

i. Due to high cost of seed, the initial investment becomes very high.

ii. Safed musli is requiring more labour and due to labour intensive, the cost of cultivation become high.

iii. Its cultivation is only possible in fertile and irrigation conditions.

iv. Lack of high yielding and quality planting material.

v. Lack of scientific knowledge regarding the package and practices of cultivation of safed musli.

vi. Unavailability of proper marketing facility.

Suggestion

Create the awareness for Safed musli cultivation in progressive farmers.

Provide financial assistance for the cultivation of crop.

Govt. creates a marketing channel from grower to consumers.

Govt. provides technical knowledge about cultivation through field demonstration, literature and exhibition etc.

Provide improved planting material by research institutes.

MARKETING SYSTEM OF MEDICINAL AND AROMATIC PLANT PRODUCTS

SHAILESH CHOUBEY
Reader and Head
Department of Economics, R.D.V.V., Jabalpur M.P.

India has one of the oldest, richest and most diverse cultural traditions associated with the use of medicinal plants. The remarkable fact is that it is still a living traditions. The knowledge of Indian people about plants and plant products not based on the application of Western categories of knowledge and pproaches like chemistry and pharmacology.

The medicinal plant industry is on the verge of entering into a high growth phase particularly in plants required for production of Ayurvedic medicines. Since the level of technologies are not very high, the progress of the industry will depend entirely on the pace at which the market lifts the material as compared to the speed with which the plant grows either in the wild or by efforts of cultivation. So, an important problem of Indian medicinal and Aromatic plant products concerns the marketing of these produce.

Marketing is a process whereby the producer and the buyer are brought together. The producer refers to all those engaged in farming land, small, tribal farmers. The buyers are classified under three main categories ;

Consumers who buy for their direct consumption producers who by medicinal plants for use as intermediate goods for processing/manufacturing, and traders who buy these products, raw or processed, for sale in the domestic market and for export. The marketing of medicinal and Aromatic plants involves several activities such as collecting and storage, their transportation to sale points/mandies, their gradation, settlement of bargains etc. In facilitating the sale-purchase of these products, a marketing system renders valuble service to an economy on several counts.

The existence of a good marketing system is an assurance to the farmer that his produce will be taken off the field. Equally importantly, such a system will indicate to the farmers the type of goods for which demand exists and the quantities in which these goods are needed. A suitable marketing system will thus not only complete the process of production by providing a link between the producer and the buyer, it will fulfill the needs of the farmers and the buyers.

A good marketing system is helpful in the development of industry. One a proper marketing arrangement can organize surpluses from this sector. Two, a marketing system in its role in a passive form (i.e. as provider of trade channels/distributive agencies) can help much if the cost per unit of the traded product is the least or minimum. This has relevance for both the existing products with existing demands and for the new products with new demands. This stimulates earnings and surpluses in the sphere of production, which in turn act as incentives to producers. Three marketing, playing an active role, can be so designed (through a set of organizations and practices) that it helps in maximizing the rate of growth of the sector. In this frame it offers attractive/profitable opportunities to producers to produce a variety of products. Four the profitable opportunities also increase tremendously as the market can bring in the foreign demands at the doorsteps of the domestic producers. Five the impulses of progressive industry can also be transmitted more rapidly through a well integrated marked set-up. Six A Well-designed marketing set-up is of considerable help in the promotion of the well-being of the people.

As described above, there are several important advantages of a marketing system. However, these do not accrue to the country, as its marketing system is beset with serious shortcomings

1. Inadequate storage
2. Inadequate means of transport
3. Inadequate facilities of credit
4. Lack of grading and standardization
5. Non availability of market information.
6. Malpractices of buying and selling
7. Inadequacy of institutional marketing.

Evil Consequences

1. Meagre earnings
2. Little incentive
3. Low marketable surplus
4. Obstructs innovative farming
5. Difficulties for planning

Improvements and progress

1. Cooperative marketing
2. Regulated markets
3. Godowns and storage facilities
4. Grading and standardization
5. Availability of credit
6. Efficient transport arrangement
7. Stabilization of prices
8. Arrangements for market information
9. Market inspection, research and training

The above suggested measures will be of great help in reforming the marketing system. As described above, some necessary steps along these lines are being taken. Also, it is imperative that this package of programmes be speedily extended to cover all medicinal goods and all farmers.

MARKET CHANNEL FOR CULTIVATED MEDICINAL PLANTS

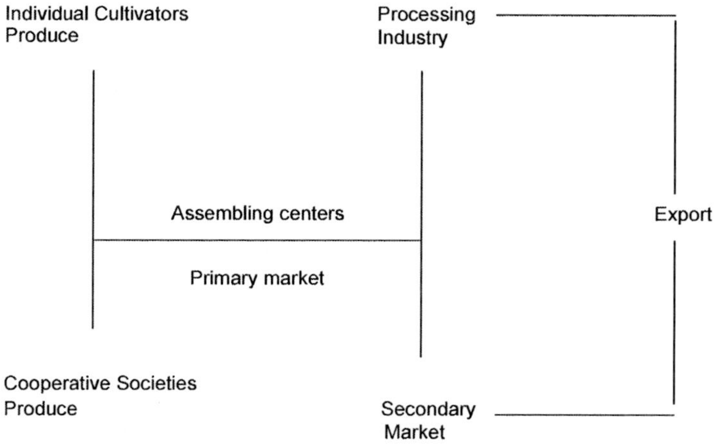

Even though the relative importance of medicinal plants may be decreasing in world trade, it is increasing in absolute terms due to renewed world wide interest in natural medicines and newer uses in food and cosmetic industries. Hence, in the medium term, prospects appear to be good.

Because of India's geographical and climatical conditions, it is in a position to export almost all plants and derivatives. The future prospects for export of medicinal plants appear to be fairly good. In order to maintain and improve upon our share in world trade in medicinal plants, it is imperative that we hone our marketing skills with suitable strategies capable of catering to the dynamic market requirements, duly supported by appropriate government policy.

An integrated approach to promote the export of medicinal plants from India right from the stage of cultivation, collection, storage, processing and marketing in an organized manner with the help of leading Indian pharmaceutical companies is suggested for a few select items beginning with the example of Isabgol.

INSECTISIDAL PROPERTIES OF MEDICINAL PLANTS

V K AGRAWAL
Professor and Head
Deptt. of Plant Physiology, JNKVV, Jabalpur (MP)

Indiscriminate use of chemicals and pesticides have resulted in several risks and ill effects such as health hazards, ecological imbalances, pest resistance, resurgence of minor pests and environ mental pollution. Intensive agriculture with the use of agro-chemicals in large quantities has no doubt has resulted in many fold increase in productivity of farm commodities, but the adverse effect of these chemicals are clearly visible on soil structure, soil microflora, quality of water, food, fodder and food materials. The W.H.O. has estimated that globally at least three million people are poisoned by pesticides every year and out of which 20,000 die.

Future agriculture warrants on ecofriendly farming which should create an ecological balance and microenvironment suitable for health and growth of soil microflora, plants, animals, farm workers and ultimately the vast population, which consume the farm produce.

The use of pesticides of plant origin for the control of pests in agriculture has a long history. Centuries before, synthetic pesticides have been developed; farmer in India protected their crops with biopesticides. The plant kingdom is a vast source of bioactive chemicals. The bioactive chemicals are safe to man and environment possesses broader scope as pest control agents as they are easily biodegradable, harmless to nature and inhibit insecticide-detoxifying enzymes in insects.

A wider range of bio-extracts are proved to be more effective in controlling many insect pests and diseases. Farmers in their traditional wisdom have identified and used a variety of plant products and extracts for pest control, especially in storage. As many as 2121 plant species are reported to possess pest management properties. At about 1005 species of plants exhibiting insecticidal properties, 297 with repellent properties, 27 with attractant properties and31 with growth inhibiting properties have been identified.

Botanical pesticides

Botanical pesticides are those pesticides, which are based on the plant materials and are known as botanical or phytochemical pesticides. Botanical pesticides are effective, commercially viable and inexpensive.

The pesticidal value of few medicinal plants are briefly reviewed below:

1. Buch or Sweet flag (*Acorus calamus* Linn.)

It is a semi-aquatic rhizomatous perennial herb belonging to the family ***Araceae***, and is cultivated for its rhizome. The dried rhizome is powdered and used for destruction of bedbugs, moths, lice and insect pests. Ether extract of rhizome shows mutagenic properties. The rhizome yield oil containing asarone and other ingredients (Jacobson, 1983)

2. Adusa (*Adhatoda zeylanica* Medicus.)

This is an evergreen perennial shrub of family ***Acanthaceae***, which is distributed throughout India. The leaf infusion of *A. Zeylanica* is used against white ants and red spiders of tea. The major components of the essential oil from leaves are limonene and an alkaloid vascine (Singh *et al.* 1965).

3. Siris tree (*Albizia lebbeck* Benth.)

This is a large, erect, unarmed spreading tree belonging to the family ***Leguminoseae*** (sub family ***Mimosaceae***), is common all over India including Andaman and Nicobar islands. The seeds petioles, leaflets and pods are toxic to cotton stainer bug adults. Bark , root and seeds are toxic to melon worm larvae. Leaves contain caffeic acid, alkaloids, kaempferol and quercetin and bark contain tannins (WOI, 1985).

4. Greater galangal (*Alpinia galanga* Swartz.)

The Greater galangal is also known as Java galangal belonging to the family **Zingiberaceae** is a herbaceous plant grown throughout India. The plant is cultivated for its rhizome in tropical areas of South East India. The essential oil contained in the rhizome shows insecticidal properties and is reported to have high knockdown action against houseflies. The essential oil contains methyl cinnamate, cinerol and camphor (Singh *et al.*, 1965).

5. Garlic (*Allium sativum* Linn.)

This herbaceous plant belonging to the family ***Liliaceae*** is believed to be indigenous to the mountain regions of Central Asia. Since ancient times, it has been known in China and Japan. Presently, garlic is cultivated in the tropical sub-tropical and temperate regions of the world. Almost all the plant parts are used as pesticide, which do not cause any adverse side effects to man and environment. The whole plant, bulbs, leaves and flowers are used either fresh or dried (Vijayalakshmi *et al.*, 1996). The pesticidal activity of garlic is due to the presence of an acrid volatile oil, which contains diallyl disulphide, diallyl trisulphide and sulphoxides derived from allicin present in it. Garlic oil is hundred percent effective on several species of mosquitoes, houseflies and major insect pests (Amonkar and Benerji, 1971). A mixture of garlic and neem oil suitable in the control of coconut mite *Aceria* (Eriophis) *garironis*.

6. Indian Neem tree (*Azadirachta indica* A. Juss.)

A large ever green tree also known as neem, margose tree or Indian lilac of family **Meliaceae** is receiving global attention as a pesticidal plant for the last two decades. The azadirachtin (Rembold, 1989) present in fruit, kernel, oil cake and leaves act as repellent, growth retardant and oviposition inhibitor when insects are treated with water extracts. The most biologically active constituent of the neem is the highly oxygenated azadirachtin A. Neem products with or with out azadirachtin have shown excellent pest control activity against infestation by hundreds of insect pests and infection by nematodes, viruses and some pathogenic fungi (Devkumar and Parmar, 1993). Neem oil is used in soaps, toothpastes (germ killer), hairtonic for killing lice (Parasite killer). Neem kernel suspension (1%) is a household insecticide as it is a mosquito and insect repellent (Pashpangadan et al., 1993, Mariappan, 1995). Pesticide formulations containing azadirachtin are now available in India, USA, Canada, Australia and Germany. The name of some of commercial botanical pesticides obtained form Neem are:

1. Margocides 2. Achook 3. Neembecidine 4. Bioneem 5. Nimbasol 6. Neeazal F, 7. Rachshak 8. Neem Gold 9. Neemark 10. Sakirina.

7. Maha Neem tree (Melia *azadirachta* Linn.)

It is a moderate sized deciduous tree found in sub Himalayan tracts belonging to the family ***Meliaceae***. Leaves protect books and woolen clothes. Leaf extracts are used against grasshoppers and locusts. Alcohol and petroleum ether extracts of stem bark of M. *azadirachta* is toxic to carpet beetle larvae (Watt. And Breyer, 1962). Semi purified extract of this plant exhibited anti feeding, toxic, growth regulating, sterilizing and ovicidal effects on a number of insects. As in Neem, Melia is known to synthesize tetra and pentanortriter-pinoids responsible for its bioactivity against insects.

8. Tobacco (Nicotiana *tobacum* Linn.)

Tobaco is an erect glandular and pubescent herb belonging to the family ***Solanaceae***, cultivated in India for its leaves. Tobacco powder andextracts are generally used to eradicate lice, and ticks in animals. The active constituent in this plant is nicotine. Sulphate of nicotine is widely used as insecticide. Other alkaloids present in tobacco are anabasine and nornicotine (Schmeltz, 1971).

9. Pyrethrum (*Chrysanthemum cinerariaefolium*)

The flower of pyrethrum is the source of pyrethrins I and II sesquiterpines, flabonoids, carotinoids and is the safest insecticide. It is especially recommended for the control of household insects like houseflies, mosquitoes, cockroaches and also for protection of food grains and edible material. It is a cultivated herb which belongs to family *Asteraceae.*

10. Cinchona bark (*Cinchona officinalis* Linn.)

Several species of this small tree are cultivated throughout India for extracting quinine. The important alkaloids present are quinine, quinidine, cinchonine and cinchonidine (Blackwell 1990) which are useful for the treatment of malaria. Cinchona alkaloids are used as preservative for fur, feather, wool, felt and textiles.

11. Karanj (*Pongamia pinnata*)

Pongamia extract of leaves and kernal serve as natural pesticides. They can be used for controlling a variety of pests. This medium sized semi-ever green tree belongs to family ***Fabaceae*** and found throughout India. Chemical constituents of pongamia are ponaglabrone, dixetonepongamol, glabrin, karanjin, pongapin and kanjone.

Pongamia oil in combination of oil extract or pongamia Aloe and Neem extracts effectively controlled a large variety of pests.

12. Tulsi (*Ocimum spp.*)

Tulsi belongs to family ***Lamiaceae*** and mainly contains citral, linalool, geranial , methyl chavicol, ocimene, eugenol etc. Use of Tulsi for pest control has a long history. It has been traditionally used to repel mosquitoes. It has both repellent and herbicidal properties. Various extracts can be prepared from dried and powdered leaves, flowers and entire plants but their effectiveness appears to vary.

13. Aloe (*Aloe barbadensis*)

It is a perennial plant with fleshy leaves belonging to the family ***Liliaceae***. The leaves contain barbolin, chrysophenol, glycoside and aglycone, aloe-emodin etc. aloe has antifungal antibacterial property when mixed with castor cake.aloe in combination with other plant products also control a variety of pests efficiently.

14. Custard apple (*Annona spp.*)

Custard apple or Sitaphal is a small tree which belongs to family ***Annonaceae*** and found throughout India in deciduous forests.The seeds of custard apple possess insecticidal and anti-feedant properties. The powdered seeds are used to destroy worms; seeds are reported to be contact poison to flies, aphids and several beetles. The combination of seed extract of neem and custard apple is toxic to housefly. it is also used as a nematicide. Different alkaloids namely annonine, municine artabortine etc. are reported to be found in custardapple.

15. Nirgundi (*Vitex nirgundo*)

It is an aromatic large shrub or small tree of about three meters in height belonging to family ***Verbenaceae***. Leaves contain two alkaloids nishindine and hydrocotylene. Thevitex of leaves are reported to possess insecticidal properties. They are used in storage godown to ward off insects. Extracts of leaves and twig show antibacterial activities. Vitex , in combination with other plant extracts prove to be successful in controlling certain pests. Vitex leaf extract in combination with Aloe , pongamia, castor and calotropis successfully control variety of pests.

16. Kiramar *(Aristolachia indica)*

It is a perennial herb belonging to family Aristolochiaceae and found throughout India. The leaves and fruits contain myristic palmitic, stearic, lignoceric oleic, and Aristolochic acid etc. It possess anti fungal, anti bacterial property. Root yield Aristolochic acid D, Aristolamide and Aristolochine.

17. Mahua (*Madhuca indica*)

It is a large tree with a dense spreading crown with dark grey or brownish leaves and scaly bark belonging to the family **Sapataceae**. Seed contains saponines, leaves gave b carotine, sitoseterol and fruit gave triterpinodes. It also contains Myristic, palmitic, stearic, Arachidic, oleic and Linoleic acid. It has insecticidal and anti-bacterial properties and controls a wide variety of pests.

18. Salvia or garden sedge (Salvia officinalis)

It is hardy shrub belonging to family Lamiaceae and found throughout India, Leaves contain flavanes and root contains diterpine quinone, The essential oil has insecticidal properties.

19. Datura (*Datura stramonium* L.)

Datura is a herb and belongs to family *solanaceae*. Plant extract is insecticidal.it contains many alkaloids such as hyoscyamine, hyosyne and atropine. Several other tropane, alkaloids like skimmianine, apophyoscine, apoatropine, tropine Ø and ß belladonine and 2 – 6 – dihydroxytropane.

20. Vividang (*Embelia ribes*)

Vividang is a shrub, belongs to family myrsinaceae. Its berries contain quinonesembelin repanone, homoembelin, homorapanone and vilangin. Berries possess insecticidal and antibacterial property. It controls wide variety of pest and many bacterial diseases.

21. Worm weed (*Artemisia nilagirica*)

Worm weed is a scented herb and belongs to family Asteraceae. Its essential oil possess larvicidal, insecticidal, antimicrobial and antifungal properties. Chief constituents of essential oil are camphor, 1 : 8 – cineole and borneol. It also contain in small quantity ß – tuzone, ß - pinene.

Thus plant kingdom is a vast source of active chemicals, which they synthesize to protect themselves from colonization of insects and other herbivores. These chemicals act as direct toxicant, repellents, behavioural and physiological disruptant. Since men become aware of the problems caused by synthetic chemicals and are looking for more safer plant protection strategies including the use of Botanicals or Phytopesticides.

MEDICINAL PROPERTIES OF WEED FLORA ASSOCIATED WITH *KHARIF* AND *RABI* SEASON

NISHA MEHRA

Deptt. of Plant Physiology, J.N.K.V.V., Jabalpur- 482004

India has great variation in agroclimatic situations, which favours the cultivation of various crops. But by the same taken, we also have in the country numerous, widely weed flora associated with *Kharif* and *Rabi* season. It is called that "Weed is a plant that is extremely noxious, useless, unwanted or poisonous", but we can not say that every unwanted or poisonous plant is weed. There are so many weeds like, *Datura, Argemone maxicana, tribulus terestris* are no doubt poisonous in nature but also having medicinal properties. Some short duration weeds like *Phyllanthus niruri* called as Jaundice plant due to having property of reducing liver disorder, weeds like; *chenopodium album, Amaranthus virdis,* and *Portulaca spp*, are rich sources of proteins, carbohydrate and minerals and we take as leafy vegetable. Owing to poor Knowledge former destroy the weeds and throw away from the field.

Therefore, the Medicinal weeds need to identification, conservation and propagation. Some of the weeds like; *Datura, Argemone maxicana, tribulus terestris* can be conserve in uncultivated land with their natural conditions. Enlightened with medicinal potential of certain weeds, some authors tend to *define weeds as plants for which uses are yet to be discovered.* But this is no way lowers the undesirable nature of weeds in agricultures or reduce our concern towards them as harmful plants.

Here giving brief idea about medicinal properties of weeds: -

Weeds found in *Kharif* season –

***Abutilon indicum* (Linn)**

Family: malvaceae

English name: Indian Mallow

Hindi name: Kakahi, Kanghi

Parts-used: Root bark, Leaves and seeds.

Description: A perennial Softly tomentose shrub, 2-3 meter high, Leaves dentate & ovate cordate Flowers-yellow, fruits schizocarpus caprule.Found in uncultivated land.

Taste: Sweet.

Chemical constituents: Plant Contains:-Gossypetin, Glucosides, Cyanidin, Rutinosides, Essenlial Oil; Root Contains: Fatty acids, Aspargin, Rafinose Stem & Leaves Contains Alkanol, Beta-sito sterol, Vanillic acid, Caffeicacid, Fumaric acid, Fructose, Glucose, Galactose, Leucin, Histidine, Threonin, Serine, Glutamic acid and Aspartic acid.

Actions: Analgesic, Laxative Expectorant, Diuretic, Carminative, Aphrodisiac, Diphoretic, Anthelmintic, febrifuge.

Used-in- Seeds: In piles, chronic cystitis, Gonorrhore; Leaves: Locally applied to boiles, ulcers, formation to painful parts; **Decoction of Leaves**: In Toothache, Gums; Given internally for bladder in flammation, **Root**: In Fever & Chest affections.

***Achyranthes aspera* (Linn)**

Family: Amaranthaceae

English name: Prickly-chaff Flower

Hindi name: Latjeara/chirchita

Parts-used: Whole plant, Leaves, Roots Seeds

Description: An annual erect herb with square stem, 1-3 feet high, yello wish pink Flowers.

Taste: Bitter and Acrid

Chemical constituents: Seeds Contain: Saponin A & B Ecdy steron, Oligosacharide, Root contains: Oleonilic acid, Maslinic acid, Unripe Fruits Contain: Saponin C & D Plant Contains: Nitrogenous Componds- Achranthine and Alkaloids.

Actions: Purgative, Diuretic, Antibiotic, Astringent, Stomachic, Digestive Expectorant, Anthelmintic, Antidotal, Emetic.

Used-in-Seeds: in Hydrophobia, Cough, Cold, Abdominal colic, Toothache, Scabies, Rheumatism, Piles, Boils, Dysentery, Snakebite, Hypotensive & Spasmogenic, Skindisease (Leprosy).

Asteracantha longifolia

Family: Acanthaceae.

English Name: Long Leaved Barleria

Hindi Name: Talmakhana

Parts-Used: Roots, Leaves, Seeds.

Description: Aspiny herb growing near moist places. Spine are yellow in whorl of Six. Leaves sessile, in whorl 6 leaves at each node Flowers bluish purple.

Taste: Bitter.

Chemical constituents: Seeds Contains:- Yellow Coloured Oil; Diastase, Lipase, Protease, Salts of Potassium; Mucilage.

Actions: Cooling, Diuretic, Sweet, Spermopoietic, Galactogogue, Tonic.

Used-in-Dropsy, Disuria, Cystitis, Jaundice, Gout, Oedema, Liver disorder, Calculii, Eye trouble

Boerhaavia diffusa **(Linn).**

Family: Nyctaginaceae

English Name: Hogweed Horsse Purslene.

Parts-used: Fresh whole plant, Root, Leaves and Flowers.

Description: A small perennial Creeping herb with purplish stem and divaricate branches and branches are swollen at nodes.

Taste: Bitter

Chemical constituents: Plant Contains: Alkaloids-Punarnavan, Sterol, Betasitosterol, Alpha-2 sitosterol, Elentine, Stearicacid, Palmiticacid, Minerals: Potassium nitrate, Sodium sulphate and Chloride.

Actions: Laxative, Stomachic, Stimulant, Anti-inflammatory Antibacterial & Antiviral, Diphoretic.

Used-in-Asthma, Anaemia, Jaundice, Scanty urine, internal inflammation, Snake venam, Epilepsy, Heart disease, Eye-troubles, Cough Bronchitis, Mycocardial Stimulant.

***Cassia occidentalis* (Linn).**

Family: Caesalpinaceae (Leguminosae)

English Name: The negro coffee.

Hindi Name - Kasondi

Parts-used- Whole plant, Root, Stem, Leaves and Fruits.

Description : A large woody herb, offensive odorous, 3-6 feet high, Compound leaves, darkgreen. Leaflets opposite, ovate-oblong or lanceolate acuminate. Yellow Flowers. Pod flate, falcate, explate between seed.

Taste: Bitter. Sweet.

Chemical constituents: Plant Contains, Emodin, Tannic acid, Fatty oils, Mucilage Amminoacids chrysophanol, Sitosterol, Xanthon, Cassiodin, Nitrogen bases-Betain, Choline, Stachvdrine, Trigonelline; Seeds Contain: Galactomanon.

Actions: Aphrodisiac, Stomachic, Diuretic, Appitizer, Anti pyretic, Anti leprotic, Expectorent, Digestive, Anti cough.Thermogenic, blood purifier, purgative.

Used-in- Cough, increases digestive power, clears throat obstruction and Decoction of Root bark in: Diysuria and inflammation. Seeds are used in Skin disorder (Leprosy)

Cyperus rotundus

Family: Cyperaceae

English Name: Nutgrass. Hindi - Nagarmotha

Parts-used: Black-Tuberous Roots/Rhizome.

Description: A perennial glabrous herbaceous sedge, leaves arisefrom under ground rhizome.

Taste- Acrid – bitter.

Chemical Constituents: Rhizome Contains: Essential oil which contains: Penene, cineol; Fattyoil which contains: Linoleic acid, Oleic acid, Myristic acid, Stearic acid; Glycerol Glycoside Cyperene, Cyperenone, Sugars Gum and Carbohydrates; Leaves Contain: Luteolin.

Actions: Astringent, Stomachic, Digestive, Blood purifier, Anthelmintic Diuratic, revulsive galactogogue, nervine tonic, anti-inflammatory, carminative, expectorant.

Used-in-Dyspepsia, Sprue, Skin disease, Diarrhoea, Thirst, Fever, Galactogogue, Uterus, Stimulant, Wound Healer Promote, Hairs growth, Anorexia, Flatulence, Diarrhoea, Epilepsy, Caliculi.

Datura metal

Family: Solanaceae

English Name: Thorne apple. Hindi - Dhatura

Parts-used-Roots, Leaves, Seeds, Fruits and Flowers.

Description: A small shrub, 3-6 feet high branches greenish white, Leaves simple dentate ovaye, glabrous, alternate or opposite; petals white but usnally purple outside, fruit with numberous prickles and are purple in colour.

Taste: Bitter

Chemical constituents: Seeds and Flowers Contain: Daturadiol, Daturalone, Fastusine, Beta-Sitosterol, Hyosine, Hyoscyamine, Atropine, Fastudin, Fastunine, Fastusic acid, Fastusidine, Fastusinine, Daluranolone, Citrostadienol, obtusifoliol, Allantoin, Caproic, Beta-Linolic, Linoleic acid, oteicacid, Datumetine, Datumetelin, Vitamin-C Chlorogenic acid and Fixed oil.

Actions: Antiseptic, Antipyretic, Narcotic, Analgesic, Spasmolytic, Anticough, Anti-inflammatory, intoxicant, emetic.

Used-in-Ulcer, Diarrhoea, Skin disease, Scabies, Pruritis, Fever, Insanity, oedema Antispasmodic, It causes vertigo but useful in Bronchial Asthma, neuralgia, myalgia hyperacidity, renal colic, caliculi.

***Eclipta alba* (Hassk)**

Family: Asteraceae

English Name: Trailing Eclipta. Hindi - Bhangra

Parts-used-Whole plant, Leaves, Seeds and Roots.

Description: A small annual and procumbent & Coarse herb, prostrating, leaves opposite, simple variable, Flowers white in Capitulum inflorescence.

Taste: Bitter/Astringent.

Chemical constituents: Plant Contains: Glucosides-Phytosterol-A, Beta-Amyrin, Wedelolactone, Luteoline, Triterpenic acid, Fatty acids, Palmitic acid, Stearic acid, Oleic acid, Linolic acid, Nicotine, Ampliptene alkaloid, Ecliptine and Resin.

Actions: Emetic, Analgesis, Spasmogenic, Antiseptic, Antiviral, Hepatoprotective Stomachic, Digestive, Carminative, anti-inflammatory, anthelmintic, Hoematinic.

Used-in- Jaundice, Ulcer, Wounds, Healing of fracture, Anti-leprotic, Cancer, Chronic fever, Elephantitis, Leucoderma, blackening of hairs, useful in digestion, Liver troubles, Anaemia, Cough, Asthma, hypertension, Strangury.

Euphorbia hirta

Family: Euphorbiaceae.

English Name: Milk weed

Parts-used-Whole plant

Description: A small annual prostrate herb with ascending and tetragonal branches, hispid with yellow hairs.

Taste: Bitter.

Chemical Constituents: Plant Contains: Quercetin, Triacentane, Ellagic acid, Phenolic substance, Euphosterol, Phytosterol, Phytosterolin, Gallic, Mellistic, Palmitic, Oleic, Linoleic acids; Zanthorhamnin wax, cautchouc, Resin, Tannins, Sugar, Mucilage, Carbohydrates, Albuminoids, Catchin, Oxalate.

Actions: Astringent, Expectorant, Lactagogue, Diuretic.

Used-in-Intestinal worms in children, Bowel complaints, Cough, useful to nursing mothers When the milk formulation is stops or deficient also useful in Leucoderma, Bronchial Asthma, Blood dysentery & Colic.

Phyllanthus niruri

Family: Euphorbiaceae

English Name: Bhuiaonla. **Hindi-**Tamlki

Parts-used- Roots, Fresh Leaves and Whole Plant.

Description: A small annual herb 10 to 30 cm. High. Flowers small greenish yellow in axils, Fruits are smell like Anola.

Taste: Bitter/Astringent.

Chemical Constituents: Leaves Contain: Bitter Phyllanthin & Hypophylanthin, Alkaloid, Lignas, Vitamin-C, Quercetin, Astrogalin, Querscitrin, Rutin; Roots Contain:

Glyco flavon; Seeds Contain: Oil which Contains-Ricinoleic acid, Linoleic acid & Linolenic acid.

Actions: Anti-pyretic, Antiseptic, Astringent, Cooling, Diuretic, Digestive, stomachic.

Used-in-Dropsy, Urinary disease, Burning sensation, Polyuria, Diarrhoea, Gastrointestinal troubles, Colic Dysentery, Dyspepsia, Jaundice, Gonorrhoea, Menorrhagia; **Leaves:** Diuretic, Mixed with salt: applied locally on Skin disease and Ulcers: Cough & Asthma (Decoction of whole plant)

Sida cordifolia

Family: Malvaceae

English Name: Country Mallow. Hindi - Bariyari

Parts-used-Roots, Leaves, Seeds and Whole plant.

Description: An errect annual or perennial herb, 2 to 4 feet high, branched, pubescent. Leaves are heart shaped, ovate, cordate, nerrate hairy with crenate margin. Flower small yellow, axillary.

Taste: Extremely bitter.

Chemical constituents: Roots Contains:- Beta-sitostarol, Alkaloids-Vasicinone, Vasicine, Vasicinol, Choline and Betain, Seeds Contain:- Ephedrine, Fatty oils, Steroids, Resin, Mucin, Resin acid, Sterculic acid, Malvalic acid, Linoleic acid, Potassium Nitrate.

Actions: Tonic, Diuretic, Astringant, Aphrodisiac, Blood purifier, Wound healing.

Used-in-Whole plant:- Juice mixed with water is beneficial in spermatorrhoea, Seeds used in: Colic & Gonorrhoea, Leaves infusion is Febrifuge, Roots are Tonic, Astringent Diuretic. **Infusion of Roots is used in:** Nervous and Urinary diseases, Cystitis, Bleeding piles, Chronic dysentery, Leucorrhoea.

***Vernonia cinerea* (Linn).**

Family: Asteraceae

English Name: Fleabane. Hindi- Sahdei.

Parts-used-Roots, Flowers, Seeds and Whole plant.

Description: A small and hairy annual herb stem erect or prostrate leaves ovate or lanceolate, toothed. Flowers violet-pinkish in capitulum (Heads.) flat topped, involucre bracts linear.

Taste: Bitter

Chemical constituents: Whole plant contains:- Lupeol acetate, Alpha-amyrin acetate, Glucoside, Quercetin, Beta-Sitosterol, Stigma sterol, Alpha-spina sterol, Potassium chloride, Alpha-amyrin, Lupeol, Flowers Contain:- Luteoline, Iso-orientin, Chrysoeriol.

Actions: Diphoretici, Anthel mantic, Anti-inflammatory.

Used-in-Roots useful in: Dropsy, Malarial fever; Flower useful in ophthalmic disease like Conjuctevitis Seeds are useful to cure worms infestation.

Weeds found in *Rabi* Season –

***Amaranthus spinosus* (Linn).**

Family: Amaranthaceae

English Name: Prickly Amaranthus. **Hindi** – Kateli Chaulai

Parts-used-Roots and Leaves.

Description: A small annual, erect herb with reddish much branched stem, axillary spines. Leaves alternate, Ovate, entire. Sharp pointed. Flowers in axillary clusters or terminal spikes.

Taste: Sweet.

Chemical constituents: Plant Contains: Carbohydrates, Proteins, Mineral, Iron, Vitamin- A and Calcium (Leaves).

Actions: Cooling, Appitizer, Stomachic, Antidote, Blood purifier, Antibillious, Diuretic, Galactogogue.

Used-in-Plant is Antidote, Antibilious, Antifatigus; useful in Inflammation, Blood impurities, Insanity, Hemorrhagic disease, Fever, Cough, Diarrhoea; Lucorrhoea, Gonorrhoea, Metrorrhagia.

***Argemone maxicana* (Linn).**

Family: Papavaraceae

English Name: Maxican Poppy / Prickly poppy. **Hindi** - Satyanashi

Parts-used-Roots and Yellow latex.

Description: A Prickly herb with spreading branches and yellow sap. Leaves Spiny & Variegated; Flowers-Yellow. Spineending, amplexicaule, lobect.

Taste: Bitter.

Chemical constituents: Roots Contain;- Berberine, Protopine nitrate, Fatty acids, Ceryl alcohol, Allocryptopine; Flowers Cantain: Tannin, Resin and Argemone oil; Root Cantain: Betasitosterol, Leaves Cantain: Glucosides, Kampferin, Anthraquinone, Essential oil & Chrysophanic acid; Seeds Contain: Tasteless oil of dark brown coluor and Carbohydrates, Albumin, Phosphate & Sulphate.

Actions: Expectorent Anthelmintic, Laxative, Demulcent, Antileprotic, Antiinflammatory, Rejuvinator.

Used-in- yellow Latex used in Jaundice, skin affection, Malaria, applied externally to stop bleeding from cuts and wounds; Root's juice mixed with peppar is given to children for the disease of supposed to be caused by cats, in woud healing, Leprose, Blood purifier.

***Calotropis procera* (Ait).**

Family: Asclepiadaceae

English Name: Shallow wort. **Hindi** - Aak/Akua/ Madar

Parts-used- Root, Leaves, Flowers and Latex.

Description : An erect shrub, young parts clothed with white cottony pubescence, Latex (Milky) flowers in umbel cymes.

Taste: Slightly Bitter

Chemical constituents: Plant Contains:- Cardiac Steroid glycoside-terpenoids. Resin, Glycosides, Carbohycraties, Aluminum, Iron, Magnessium, Sodium.

Actions: Analgesic, Resolvent, Anti-inflammatory, Emetic, Expectorant, Stomachic, Digestive, Laxative, Anthelmintic, Depurative.

Used-in- Skin diseases, Leucoderma, Eczema, Ulcer, Piles, spleenomegaly, Dropsy, stranguory, Antheliminntic: Fresh Leaves in: Rheamatiod Arthritis, Inflammation; Dried & yellow Leaves Useful in cough and Asthma.

***Chenopodium album* (Linn).**

Family: Chenopodiaceae

English Name: White goose foot/Lamb's guarters. **Hindi** - Bathua

Parts-used-Leaves and Seeds.

Description: A small erect-annual herb, branches and leaves are of various colours, flowers are small green in campact spikes.

Taste: Sweet.

Chemical constituents: Plant Contains:- A volatile oil, Which contains Ascaridol; Saponin, Vitamin-C, Magnissium Phosphate, Carotene.

Actions: Anthelmintic, Sweet, Laxative, Antipyretic, Appitizer, Stomachic.

Used-in-Bleeding piles, Dysentery, For longevity of life, Intrinsic heamorhage Cough, Polyuria, Diysuria.

***Convolvulus pluricaulis* Choisy.**

Family: Convolvulaceae

English Name: Shankh Pushpi. **Hindi** - Shankhahuli

Parts-used- Whole plant, Roots, Leaves and Flowers.

Description: A small prostrate creeping herb, branches pubescent, flowers white.

Taste: Bitter

Chemical constituents: Plant Contains: Alkaloid shankhpushpin (Action compounds) Essential oil, Methly pentos, Scamonolic acid and Palmitin acid, Phytosterol, Ipuranol, Sucrose reducing sugars, starch and Olein.

Actions: Brain, Tonic, Laxative, Stomachic, Anthelminitc, Spasmolytic.

Used-in-Effective on brain and heart, positive effect on blood pressure of acteylcolonine reduce blood pressure and tension, increaces memory, Effective on Thyroid high secretion; useful in mental disease, Epilepsy, Insomnia and Venam.

***Gynandrosis gynandra* (Linn).**

Family: Capparidaceae

English Name: Pandri-tilwan. **Hindi-** Hul-Hul

Parts-used-Roots and Seeds.

Description: Annual, erect branched, hairy, pungent aromatic herb; Leaves digitately 3-5 foliate, with long petioles; White flowers in Corymbose racemes, petals stalked, bracts leafy; siliqua elongate, striated; seeds reniform or orbicular.

Taste: Acrid.

Chemical constituents: Plant Contains:- Essential oil, Seeds Contain:- Cleomin.

Actions: Seeds Stomachic, Digestive, Stimutant, Anthelmintic, Inflammatory, Roots Stimulant, Diphrotic.

Used-in-Intrinsic Haemorrhage, Asthma, Cough, Indigestion, Fever, Skin Eruptions, Leprosy, Polyuria, Dyscrasia, Genital disorder, Anaemia.

***Oxalis corniculata* (Linn).**

Family: Oxalidaceae

English Name: Indian sorrel Hindi - Tinpatiya

Parts-used-Whole plant and Leaves.

Description: A small prostrate appressed pubescent, perennial herb; leaves trifoliate with long rachis, leaflets obcordate, sessile Flowers: yellow solitary or grouped in 2-6, capsule oblong, hairy.

Taste: Sour-Sweet.

Chemical constituents: Plant Contains: Malic acid, Tarteric acid, Citric acid, Potassium oxalate.

Actions: Cooling, Refringirant, Analgesic, Appitizer, Stomachic, Digestive.

Used-in- Scurvy, Menstrual disorder, helps in fertility, Redness & Vaginal watery discharge, Stomachic, Hypotensive, Antipyretic, Andidote, Anti-inflammatory, Antibacterial, useful in Diarrhoea Piles & Fever.

***Solanum nigrum* (Linn).**

Family: Solanaceae

English Name: Black night shade. **Hindi-**Makoy

Parts-used-Whole plant, Flowers, Leaves and Fruits.

Description: An errect annual, diffused much branched herb, 1-2 feet high. Flowers small & white. Fruits berries, ripen fruits are black or orange-yellow.

Taste: Bitter / Acrid.

Chemical constituents: Plant Contains:- Alkaloids, Reducing Sugars, Glycosides, Saponins, Steroids, Leutein,

Lycopene, Vitamin-C. Glucose, Fructose, Caffeic-Solasodine, Tomatidenel, Solamargine, Solasomine Trigogenin, Potassium, Iron, Sulphur, Calcium, Phosphorus.

Actions: Anti-inflammatory, Analgesic, Diuretic, Rejuvenator, Spermopoeitic, Dumulecent, Expectorant, Anthelmintic, emollient, diphoretic, hydragogue.

Used-in- Fever, Diarrhoea, Colic, Piles, Oedema, Skin disease (Psorisis Syphilis, Prurits) Spermopoeitic, Polyuria, Anti-emetic, Useful in heart disease, Spleen & Liver megaly, Rheumatism, Dropsy, otalgio, Piccough.

***Tribulus terrestris* (Linn).**

Family: Zygophyllaceae

English Name: Land Caltrops. **Hindi** - Gokhru

Parts-used- Fruits and Roots.

Description: A small creeping herb. all parts are clothed with hairs. Leaves pinnate Compound, opposite, leaflets 4-7 pairs, oblique and hairy. Flowers small axillary of yellow colour. Fruits with five pairs of spines.

Taste: Sweet.

Chemical constituents: Fruits Contain:- Alkaloids, Fixed oil, Essential oil, Resin, Steroids, Saponins, Vitamin-C, Proteins and Nitrates.

Actions: Cooling Stomachic, Aphrodisiac, Diuretic, Tonic, Refrigerent, Diphoretic, Anti-inflammatory, Carminative, appetizer, expectorant, anodyene.

Used-in-Disuria, Nephritis, Urinary calculi, Impotency, Seminal debility, Piles, Cough, Asthma, Polyuria, Heart disease.

WEEDS –A POTENTIAL SOURCE OF LIFE SAVING DRUGS

N.T. YADURAJU. V.S.G.R. NAIDU AND ARCHANA TRIVEDI
National Research Centre for Weed Science
Maharajpur, Jabalpur-482004 (M.P)

Herbal medicines are in great demand in both developed and developing countries in primary health care because of their great efficacy and little or no side effects (Narula *et al.* 2000). The turnover of the herbal medicine in India is about US $ 1 billion with a meager export of about US $ 80 million per year, which is growing at 7% per annum (Kamboj, 2000, Nautiyal, 2002). Most of these medicinal plants are grown under natural ecosystem and are said to be weeds. In fact all unwanted plants are weeds but all weeds are not unwanted plants. Weeds are the source of several potential life saving drugs. A large number of weeds have medicinal properties and most of them are available in cultivated fields. For example *Eclipta alba, Phyllanthus niruri, Solanum nigrum* etc. are found in crop ecosystem. Owing to the poor knowledge about the utilization of these weeds, farmers throw them away from their fields. By the use of selective herbicides in the cropping systems, a group of weed species is selectively killed. This, in long run, may lead to the elimination of valuable weeds from the cropland ecosystem. There is a need to conserve the medicinal weeds by educating the farmers about the cultivation, utilization and additional revenue.

The medicinal weeds can be cultivated by fitting them into the existing cropping systems. Some of the medicinal weeds with sort growing season that coincides with the fallow period between two crops can be grown as pure crop and the others can be grown as intercrops for efficient utilization of the land and for higher net returns. Since many medicinal weeds are available in the natural ecosystem for commercial exploitation, doubts are raised about whether the cultivated produce would have viable market. Therefore there is a need to assess the total requirement, productivity potential, extent of utilization, market channels and price structure.

List of Kharif Weeds of Medicinal Importance

Botanical Name	Family	Local Name	Habitat	Flowering and Fruiting	Medicinal Uses
1	2	3	4	5	6
Ageratum conyzoides	Asteraceae	Gondhal	Crop and non	September to January	Juices of leaves are used in cut injuries to prevent bleeding. Tribal use this for number of diseases such as leprosy, piles, swelling of body, uterine disorders, and skin diseases and also for snakebites.
Ammania baccifera	Lythraceae	Dadmari	Mostly in crop land	October to December	Leaves are used in fever, skin diseases and rheumatism.
Biophytum sensitivum	Oxilidaceae	Lajvanti	Crop and non land	August to November	Leaves are antiseptic and used for healing wounds and ulcers. Plant is also useful in fever.
Boerhhavia diffusa	Nyctang-inaceae	Punarnava	Crop and	Throughout the year	Root is diuretic, purgative, laxative and also useful for anaemia.
Caesulia axillaries	Asteraceae	Bangra	Crop and non crop land	July to September	Fresh herb crushed and applied on wounds.
Cassia occidentalis	Caesalp-iniaceae	Kasaundi	Non-crop area	August to January	
Cassia toru	Caesalp-iniaceae	Chakunda	Crop and non	August to June	Leaves roots and seeds purgative. Seeds past with water is useful in skin diseases.
Cleome viscosa	Ranan-culaceae	Hur-hur	Mostly in non-	September	It is useful in all types of skin diseases. Past of root with limejuice is applied in snakebite and ringworm. Leaves useful in fever and decentry. Leaf juice is used for ear trouble. Plant is used in bronchitis. The seeds are used for stomach disorder.

1	2	3	4	5	6
Corchorus acutangularis	Tiliaceae	Koshta	Often found in crop land	August to October	The seeds powder help in removing the worms from the intestine. Poultice of seeds are effective forrheumatic joints. Root decoction is used in fever.
Cyperus rotundus	Cyperaceae	Motha	Crop and non	August to	Seeds stomachic and given in pneumonia. Leaves are used as tonic. The whole plant along with young leaves of Azadiractica indica and black pepper with leaves of ocimum sanctum is made into decoction the vapor is used to cure malaria fever decoction is given three times a day for the same purpose.
Datura metal	Solanaceae	Kala Datura	Mostly in non crop area	August to January	Tubers diuretic. Root paste is applied for healing wounds and sores.
Datura stramonium	Solanaceae	Datura	Often found in crop land	August to June	Seeds, leaves and root are given in insanity, skin diseases as antiseptic.
Eclipta alba	Asteraceae	Bhranjraj	Crop and non	Whole year	Dried leaves are applied in boil, sores, rheumatism and painful tumours. Leaves useful in bronchitis or asthma. Inhalation of burning leaves gives relief in asthma.
Euphorbia hirta	Euphor-biaceae	Choti dudhi	Crop as well in non crop land	Whole year	Used in hepatic and spleen enlargement and in skin troubles. Farmers used the herbal juice for sickle wounds. The leaf juice is used for eternal application and bacterial infection. Leaves paste is applied to cure white spot formed due to burning. The oil prepared for the herb is a great repute as hair dye and has cooling affect on the brain.
Leucas aspera	Lamiaceae	Gumma	Crop and non crop land	August to February	The plant is believed to effective remedy for disease like bron chitis, asthma and cough. The leaves posses anthelmintic prop erties and used to kill intestinal worms of children. The milky juice of plant is applied on lips against cracking.

1	2	3	4	5	6
Malvastrum coromondalinum	Malvaceae	Barirary	Mostly in non crop area	Whole year	Plant extract is given to lactating mother for increasing milk.
Phyllanthus niruri	Euphorbiaceae	Bhui amla	Mostly in crop land	July to October	The leaves are very helpful to relieve feverishness and produce coolness and soothing effect.
Rungia repens	Acanthaceae	Kharmor	Commonly in non and non	August to November	Leaf juice is useful in skin eruption and painful swelling. Flowers with honey for cough and cold of children.
Sida cordifolia	Malvaceae	Bariari	Crop and non	August to October	Plant is used in dysentery. Leaves anti-inflammatory.
Tephrosea purpurea	Fabaceae	Pit papda	Mostly in non crop area	August to October	It is a reputed drug of jaundice. Plant and the milky juice are diuretic. Bruised leaves mixed with castor oil and applied to scalp, cures fungal infection.
Trianthema monogyna	Aizoaceae	Lal subani	Crop and non crop land	August	The drug is used for treatment of rheumatism. It is also useful in neurological disorders like sciatica, headache, fever and uterine disorder.
Tridax procumbens	Asteraceae	Tal-muriya	Crop and non crop area	Whole year	The root juice is used in heart irregularity and it is also used for the healing of wounds. Leaf juice is given in dropsy.
Xanthium strumarium.	Asteraceae	Adjasos	Crop and non crop land	September to June	It is diuretic. Used in dropsy and body swelling caused by liver disorder. Crushed leaves with salt are applied on boils. It is useful in central nervous system depressant. Fruits are said to be effective in treatment of smallpox.

List of Rabi Weeds of Medicinal Importance

Botanical Name	Family	Local Name	Habitat	Flowering and Fruiting	Medicinal Uses
1	2	3	4	5	6
Abutilon indicum	Malvaceae	Kanghi	Crop and non	May to June	Decoction of leaves is used as eye/mouth wash and taken internally for stone in bladder. Decoction of sees and leaves is used in piles.
Achyranthes aspera	Amaranthaceae	Chichita	Crop and non crop land	June to September	Entire plant is medicinally used in cough, cold and enlarged spleen. Leaves juice is used to heal wounds. Plants is also used as diuretic and purgative.
Alternanthera sessilis	Amaranthaceae	Kanchari	Crop as well non crop land	August to December	It is febrifuge. Shoot contains protein and iron.
Anagallis arvensis	Scrophulariaceae	Krishneel	Crop land	September to February	Whole plant is used in laprosy, dropsy and epilepsy. It is also useful in fish poisoning and snakebite.
Anisomeles indica	Lamiaceae	Sugandhi	Crop as well non crop land	September to January	Plant carminative, astringent, tonic. Oil from plant is used in uterine infection.
Argemone mexicana	Papaveraceae	Satayanasi	Commonly in	January to	Juice diuretic, alterative, anodyne and hypnotic. Seeds are laxative, nauseate and used as antidote of snakebite. Fresh root has an anodyne effect. *Ayurveda* uses it for eye disease also. *Unani* system uses for purifying blood. Milky juice is used to relieve blisters, and heal indolent ulcers. The plant is useful in cough and cold.
Blumea lacera	Asteraceae	Kukrondha	Crop and non crop land	January to May	Plant juice is used in bleeding piles and diarrhoea. The plant is used to cure bronchitis. Root kept in mouth cures mouth diseases.

1	2	3	4	5	6
Blumea spp.	Asteraceae	Kakorandha	Non crop land, rare in crop land	January to May	The plant is used to cure bronchitis.
Chenopodium album	Chenopodiaceae	Bathua	Mostly in crop land	January to February	Root kept in mouth cures mouth diseases.
Crotolaria sericea	Papilionaceae	Ghunghiania	Mostly in non crop area	February September	Leaves laxative, iron rich, given in spleen disorder and general debility.
Cynadon dectylon	Poaceae	Doob	Crop and non crop land	February to Whole year	It is used in lowering blood presure. Plant is also used in scabies. The extract prepared from the plant along with garlic and warm mustard oil is rubbed in body to relieve body pain. Whole plant is useful in cough. It stops bleeding from cuts and wounds. The plant is also useful in bleeding piles.
Evolvulus alisinoides	Convolvulaceae	Sankhpuspi	Crop and non crop land	June to August	Roots are taken with *mishri* as memory sharpens. Locally the leaf juices dropped into nostrils to stop nose bleeding.
Lathyrus sativa	Fabaceae	Tawara	Mostly in crop land	November to January	It is used as brain tonic. Juices of whole plant or tablate made by mixing mishri are taken internally in bone fracture.
Melilotus indica	Fabaceae	Van mathi	Crop as well as non crop land	October to January	Seed-oil cathartic. Seeds toxic cause paralysis.
Merrimia emerginata	*Convolvulaceae*	Muskani	Mostly in non crop area	January Nov. to April	Seeds are used in bowel complaints and in infantile diarrhoea
Oxalis latifolia	Oxilidaceae	Khati buti	Crop as well non crop land	September to May	A decoction of plant is said to act as diuretic and alterative, useful in rheumatism. Leaf juice is given in migrane.

1	2	3	4	5	6
Portulaca oleracea	Portulaceae	Lunia bhagi	Crop as well as non crop land	October to December	Plant with cumin seeds are pounded and orally taken with water thrice a day in diarrhoea and dysentery. The past of the plant is massaged forehead to relieve headache. The fresh juice of plant is beneficial for the treatment of anemia and piles. The fresh leaf juice is mixed with cow butter is used in treatmnet of jaundice.
Rumex dentatus	Polygonaceae	Jangali palak	Crop as well as non crop land	November to March	The plant is used in disease of liver and kidney and lungs. Leaves past applied to bone and other skin diseases.
Scoparia dulsis	Scrophulariaceae	Methi patti	Generally found in non crop area	January to September	Root considered as antiseptic and astringent. Used in cutaneous disorder.
Sida acuta	Malvaceae	Bala	Crop as well as non crop land	September to January	The decoction of plants is used in eye troubles. The leaf infusion is used in fever and bronchitis. The root extraction is orally taken in diarrhoea and dysentery. The green plant is used in anemia.
Sida rhombifolia	Malvaceae	Bariyari	Mostly in non crop area	September to October	Paste prepared from leaves and young shoots applied over boils and also used as scabies and other skin diseases. Roots are used in urinary troubles. Rot extract is also given as tonic in nervous weakness. Roots are beneficial for expelling worms from alimentary canal.
Solanum nigrum	Solanaceae	Makoi	Crop as well as non crop land	November to January	The plant is used as cardiac tonic. It is important drug for the treatment of rheumatism and neurological disorders.
Solanum surrattense	Solanaceae	Katali	Mostly in non	Whole Year	The plant is regarded useful in tuberculosis and rheumatism. The pounded leaves are applied on swelling. It is anti-diabetic,

1	2	3	4	5	6
					laxative. Decoction of berries and flowers is useful in cough. The root of the plant and *Alternenthera sesilis* ground with water is given in urinary troubles.
Sphaeranthus indicus	Asteraceae	Gorakhmundi	Crop as well non crop land	January to June	The drug is used in cough, asthma, pain in chest and certain kinds of fevers. It is diuretic and considered useful in controlling of stones in bladder. Plant is used as aphrodisiac, diuretic.
	Asteraceae	Akarkara			
Spianthus acmella	Zygophyllaceae	Gokshru	Crop as well non crop land	January to June	Root and seeds used as anthelmintic. The plant improves digestion and purifies blood.
Tribulus terristris	Asteraceae	Tal-muriya	Mostly in non-crop area	June to October	The pungent flower head is chewed to relief in toothache, throat and gum infection.
Tridex procumbens			Crop and non	Whole Year	Powered herb is useful in cough and difficult breathing. It is a diuretic and used in disolving kidney stones. It cures genito-urinary disorder. Crushed leaves with salt are applied on boils.

ISSUES AND STRATEGIES INVOLVED IN CHOOSING MEDICINAL PLANTS FOR RESEARCH AND DEVELOPMENT

J.P. TIWARI

Retd. Dean College of Horticulture, Mandsaur, JNKVV, Jabalpur

Dealing with organic materials as opposed to synthetics requires more careful consideration. Processing, Trading, Marketing and Cultivating medicinal plants require holistic approach at much high level than what is envisaged for food and fiber plants. Current export potential cannot be our sole or a major criteria in choosing a medicinal plants for development. The entire export market of medicinal plants (largely from third world countries) is exposed to substitution by synthetics or tissue culture factories. Experiences with the new biotechnologies have all led to or are leading to the displacement of natural products of third world countries by cultured products or alternative plants in the developed countries. Issues that should be considered before embarking on development any medicinal plant are highlighted and list of plants which are likely to remain in demand in the fore seeable future are listed.

Enlightened businessmen and traders now see "money" in processing, trading and cultivating medicinal plants. This note is prepared primarily to assist such persons and to bring to heir notice various issues they should consider before they embark upon this venture.

Philosophy and practice of India's national Heritage, Ayurveds, is increasingly being accepted and appreciated globally by Intel lecture laymen. Locally preferred plant based medicines is also having renaissance throughout the world. However many of the therapeutically useful plants are no longer available from the wild in quantities required Non-availability of equality raw material is not only because of the population pressure and increased in demand but also because of the fact that 70% of the plants drugs involve destructive harvesting-roots (29% of the plants), rhizomes (4%), whole plants (16%), barks (14%), wood (3%) and stems (6%). (Ved et al. regulating export of endangered medicinal plant species- need for scientific rigour. Current Science 75 (4): 341-345. (1998) requirements. However, this is a complicated proposition since, for non-traditional food crops, the traditional sequence of R and D activities for technology generation (Research and Development Programme > Technology> Product> Consumer) is reversed, beginning instead with a focus on consumer: Consumer > Producer > Technology > Research and Development Programme. For medicinal and neutraceutical plants the situation is still more complicated and requires new style development strategy since for most of them, even after they are fully developed, area under cultivation will be modest and thus R and D investment per hectare or value of the crop tends to be higher.

Some of the issues to be considered while finalizing commercial cultivation and processing of medicinal plants for marketing

- *Indigenous Demand and Therapeutic Importance be Given Priority Over Export, Except Isabgol and Senna:* Export values for all plants and plant products, except Isagbgol, Opium and Senna are low, fluctuating, scanty, sometime contradictory and because of strong resentment by exporters, number of plants and their products which are banned for export is often subsequently downwardly revised, Moreover, illegal export of banned items cannot be effectively checked, Thus except Isabgol and Senna, as far as medicinal plants are concerned, export figures cannot be our sole or a major criteria in prioritizing and choosing plants for R and D activities. At the same time, it must be stated that the UK-based consultancy firm McAlpine, Thorpe and Warrier estimates the world market for plant based formulations at Rs. 51,200 crores (US $ 14.2 billion) and Dr. S.S. Handa, Director (now retired), RRL, Jammu, has stated that by the middle of the 21st century the global market of herbal drugs will be US $ 5 trillion and he wondered whether Indian companies will survive with their apathy to R and D in herbal drugs.

- Number of scientists working on medicinal plants are few, the channels of communication between scientist working on various aspects of therapeutic plants are not well organized and literature on relevant aspects of a particular medicinal plant are published in scattered journals. Before embarking on cultivation and /or development of any plant, it is highly essential to collect maximum literature and the plant be cultivated in an appropriate agro-climate and that too after consulting intelligent farmers of the same area.
- During formative yeas, area under a particular medicinal plant be restricted and expanded over the years only after gaining insight for the various relevant issues. Information on cultivation aspects on medicinal plants is meager since basic information on all of these plants is meager and number of scientists engaged in research o these plants are few.
- Chemical fertilizers may be used in raising a plant when active ingredient alone is of therapeutic importance and it is isolated for therapeutic use. However, only organic farming should be practiced in raising medicinal plants when the crude material or extract is to be used therapeutically. Often, buyers insist that only organically grown produce be utilized therapeutically and for controlling diseases and pests. While growing the plant and/ or during their storage (post-harvest handling) only biological control measures be adopted, where possible.

- *Marketing :* Strategy for marketing be worked out before raising the crop.
- *Processing:* This requires sophisticated equipments and procedures. To procure good quality raw material, possibility to pay the price to the farmers on the basis of quality be considered and details of contract cultivation be negotiated and finalized with the farmers.
- Latest literature on marketing, processing and related information on medicinal plants can also be accessed by browsing through the data on the internet and any new marketing information can also be uploaded on the internet.
- *Drying of the produce:* Leaves etc. of medicinal plants often require to be dried quickly (to have quality produce), in shade and not at very high temperature. Possibility to shade dry the produce, utilizing solar energy be explored.
- *Storage:* organic materials, as opposed to synthetics, require elaborate arrangements of storage. Possibility to store your dried produce in commercial cold storages before they are processed, be considered to minimize infestation.
- China emphasized their own traditional health care system, which is evident from the fact that traditional medicine still retains a 40% share of the medicine market nation wide. China cultivates medicinal plants in 4,60,000 hectares ant it enjoys the lion's share in worked market, They are able to achieve this supremacy by establishing a close working relationship among field scientists, pharmacologists and clinicians. In India, we have not cared that nature would stop supplying these invaluable ingredients of our essential use and have hardly tried to adopt strong policy to protect them. The situation is becoming grave with the disturbances like bio-piracy. This is one more reason why we need to

learn how to collaborate with individuals and develop inter-institutional linkages. Data on consumption and how prices of individual medicinal plants have increased over the years are not being hared by the user industries. Neither any apex body of user industries is able to provide such data- TADERS OF MEDICINAL PLANTS ARE HIGHLY SECRTIVE, Under such precarious situation, for long-term strategy (not for immediate gain) develop foundation and insight for plants which meet one or more of the following criteria:

1. It is used in crude form and is used since hundreds of years (In crude form, due to the presence of co-effectors, the therapeutic effect is often more favorable in comparison with pure substances. Moreover, if a slow release and prolonged effect is desirable, medicines of plant origin are preferred in modern medicines).
2. Efficacy of its products has proved of considerable therapeutic importance (for saponin diuretics, expectorants 7 bitter principles, preparation of pure substances are technically complicated and / or not economical).
3. Its development has long-term impact on providing genuine quality raw material for traditional systems of medicine and/or for western medicine.

Medicinal plants having high demand are recommended for cultivation by medicinal plant board (Annexure I) in 2001.

List of plants (Annexure II) recommended for study while preparing perspective plan (Vision 2000) of National Research Centre for Medicinal and Aromatic Plants were finalized, utilizing above stated criteria long with socially desirable goals/ considerations. This list was finalized after extensive discussion with: (1) Dr. Ranjit Roy Choudhary, and Eminent Scientists and Professor Emeritus, Indian Institute of Immunology, New Delhi (who has been awarded Padma Shri by the Government) after reading his books published in 1992 by WHO titled: "herbal medicine for Human Health" and (2) Dr. Sukh Deve, a world renowned natural Product Chemist. Similarly, QRT of the ICAR on All India Coordinated Research Project on Medicinal And Aromatic Plants (report submitted during Jan. 1999) has identified 57 plants (Annexure III) which have been recommended to be prioritized for germplasm collection, characterization and developing their cultivation practices.

Annexure - I

I. List of medicinal plants identified for cultivation and development by Medicinal Plants Board, New Delhi.

S.No.	Common Name	Botanical Name	Advertised
A.	**Annuals**		
1.	Isabgol	*Plantago ovata, Forsk.*	
2.	Long Pepper	*Piper Longum Linn.*	
3.	Senna	*Cassia angustifolia Vahl.*	
4.	Ashwagandha	*Withania somnifera.*	Yes
5.	Kalmegh	*Andrographis paniculata(Burm.f.)*	
6.	Safed Musli	*Chlorophytum borivillianum C. arundaenaceum.*	Yes
7.	Liquoris	*Glycyrrhiza glabra Linn.*	
8.	Bhumi amla	*Phyllanthus niruri*	Yea
B.	**Perennials**		
1.	Aonla	*Emblica officnalis Gaertn.*	
2.	Ashoka	*Saraca asoca*	
3.	Bael	*Aegle marmelos. Corr.*	
4.	Guggal	*Commiphora wightii (Arni). Bhandari.*	
5.	Sandal Wood	*Santalum album Linn.*	
6.	Kokum	*Garcinia indica Linn.*	
C.	**Climbers**		
1.	Giloe	*Tinospora cordifolia*	
2.	Glory lilly	*Gloriosa superba Linn.*	
3.	Madhunasini/ Gudmar	*Gymnema sylvestre.*	Yes
4.	Satavari	*Asparagus racemosus. Willd. ex. Royale.*	Yes
D.	**Plants for High Altitudes**		
1.	Atis	*Aconitum heterophyllum. Wall.*	Yes
2.	Chirata	*Swertia Chirayita*	Yes
3.	Indian Barberry/ Rasont	*Berberis aristata DC.*	
4.	Jatamansi	*Nardostachys grandiflora.*	

5.	Kutki	*Picrorhiza kurroa*	Yes
6.	Kuth	*Saussurea lappa*	Yes

II In Addition to the above, the following species also need attention.

1.	Patharchoor	*Coleus barbatus Benth.*	
2.	Sarpagnadha	*Rauvolfia serpentina, Benth.*	Yes
3.	Tulsi	*Ocimum sanctum*	27 Species

Medicinal Plants Board New Delhi has also advertised in Times of India (Delhi Edition) dated September 19th 2001 for contract farming arrangements between the farmers and industrialists/traders for following species.

1.	Makoy	*Solanum nigrum*	Yes
2.	Pushkarmool	*Inula racemosa*	Yes
3.	Vai vidang	*Embelia ribes Burm. F.*	Yes
4.	Vatashabh (Vish)	*Aconitum ferox/A.chasmanthum Stapf.*	Yes
5.	Brhmi	*Bacopa monniera Wettst.*	Yes.
	Total: 32 Species		

Annexure - II

Mandate

1. Isabgol (*Plantago ovata*)
2. Senna (*Cassia angustifolia*)
3. Palarosa and Lemongrass (*Cymbopogon* sp.)
4. Liquorice (*Glycyrrhiza glabra*)
5. Asgandh (*Withania somnifera*)
6. Guggal (*Commiphora wighitti*)
7. Safed musli (*Chlorophytum* sp.)
8. Aloe (barbadensis)

Additional crops for agricutlural research

1. *Asparagus racemosus* (Satavari)
2. *Tinospora cordifolia* (Guduchi)
3. *Bacopa monnieri* (jal Brahmi)
4. *Evolvulus alsinoides*(Sankhpushpi)
5. *Leptadenia reticulata* (Jivanti, Dodhi)
6. *Phyllantus amara*
7. *Silybum marianum*
8. *Andrographis paniculata* (Kalmegh)

Plants useful in traditional systems of medicine

1 *Acorus calamus*
2 *Adhatoda vasaca*
3 *Aegle marmelos*
4 *Albizzia lebeck*
5 *Artemesia annua*
6 *Azadirachta indiaca*
7 *Bacopa monniera*

21 *Hibicus rosa sinensis*
22 *Holarrhena antidyenterica*
23 *Lowsonia alba*
24 *Momordica charantia*
25 *Phyllantus amara*
26 *Picrorrhiza kurroa*
27 *Pterocarpus marsupium*

8 *Borehaavia diffusa*
9 *Butea monosperma*
10 *Cassia acutifolia*
11 *Centella Asiatica*
12 *Cinnamomum zeylanicum*
13 *Convolvulus pluricaulis*
14 *Crataeva nurvala*
15 *Cissus quadrangualris*
16 *Eclipta alba*
17 *Elettaria cardamomum*
18 *Embelia ribes*
19 *Eugenia jambolana*
20 *Gymnema sysvestra*
28 *Santalum album*
29 *Saussurea lappa*
30 *Emecarpus anacardium*
31 *Streblus asper*
32 *terminalia arjuna*
33 *Terminalia chebula*
34 *Teminalia belerica*
35 *Thymus vulgaris*
36 *Tribulus terristris*
37 *Tripteriguuym wilfordi*
38 *Urginea maritime*
39 *Vitex negundo*

Newly emerging plants having proven therapeutic importance

1. *Mappia foetida* (camptothecin; anticancer)
2. *Taxus wallichiana* (Taxol; anticancer)
3. *Croton sulyrantus* (Plaunotol; anti-gastric ulcer)
4. *Garcinia cambogia* (Hydroxycitric acid' anti-gastric ulcer)
5. *Gymnema sylvestre* (Gymnemicaida;?)
6. *Hydrastis Canadensis* (Berberine and hydrastine;?)
7. *Tylophora asthamatica* (Tylophorine; anti-asthma)
8. *Centella asiatica* (Brahminiside, etc, skin disease ?)
9. *Podophyllym emodi* (Podophyllotoxing; anticancer)
10. *Ginkgo biloba* (Ginkgolides; geratric medicine)

Kasturi Bhindi (*Abelmoschus moschatus*)

Adusa (*Adhatoda vasica* nees)

Satawar (*Asparagus spp.*)

Ulat kambal (*Abroma augusta*)

Bel (*Aegle marmelos*)

Saptparni (*Alstonia scholaris*)

Dhritkumari
(*Aloe vera*)

Kalmegh
(*Andrographis paniculata*)

(*Amorphophallus campanulatus*)

Suchi booti
(*Atropa baladonna*)

Isarmool
(*Aristolochia indica*)

Vidhara (*Argyreia nervosa*)

Brahmi (*Bacopa monnieri*)

Neem (*Azadirachta indica*)

Bajradanti (*Barlaria cristata*)

Daru Haridra (*Berberis aristata*)

Shivlingi (*Bryonia laciniosa* l.)

Bursera (*Bursera delpechimum*)

Senduri (*Bixa orellana L.*)

Sanai (*Cassia angustifolia*)

Gataran (*Caesalpinia crista*)

Bawbchi (*Carum bulbocastanum* Clarkr)

Amaltas (*Cassia fistula*)

Casondi (Cas*sia sophora*)

Sadabahar (*Catharanthus roseus*)

Mandukparni (*Centella asiatica*)

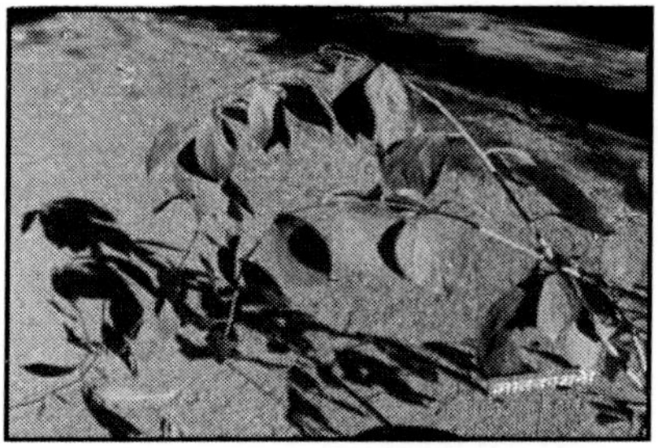

Malkangni (*Celestrus paniculata*)

Harfareowri (*Cicca acida*)

Safed Musli (*Chlorophytum borivilianum*)

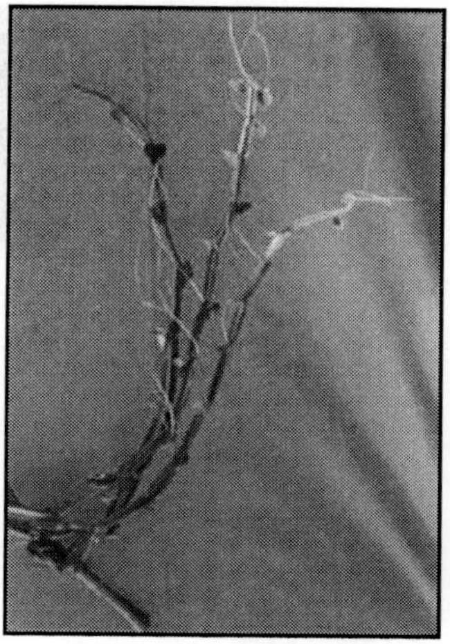

Hadjor (*Cissus quodrangularis*)

Mamira (*Coptis teeta*)

Aprajita (*Clitoria ternatea*)

Jangli ajwain (*Coleus aromatica*)

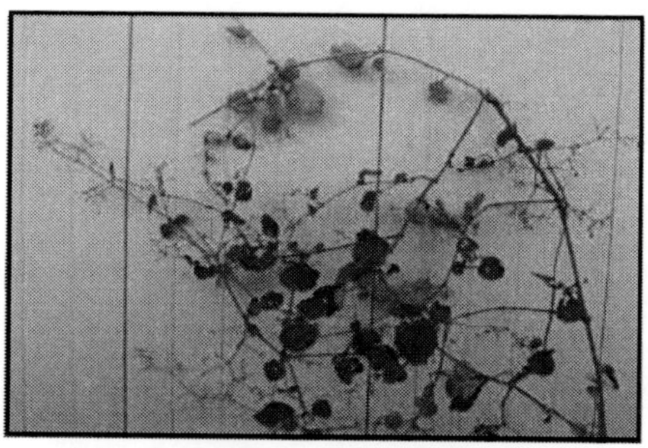

Coleous, Patharchur (*Coleous barbetum*)

Guggal
(*Commiphora wighti*)

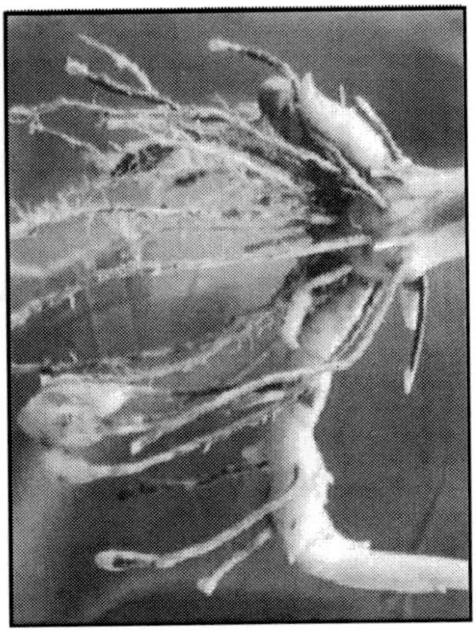

Ama Adrak (*Curcuma amada*)

Haldi (*C. aromatica*)

Deshi Haldi (*C. longa*)

Keokand (*Costus speciosus*)

Lemon grass
(*Cymbopogon flexiosus*)

Rosa grass
(*Cymbopogon martini*)

Dhatura (*Datura mental*)

Dhatura (*Datura stramonium*)

Ratalu (*Dioscorea bulbifera*)

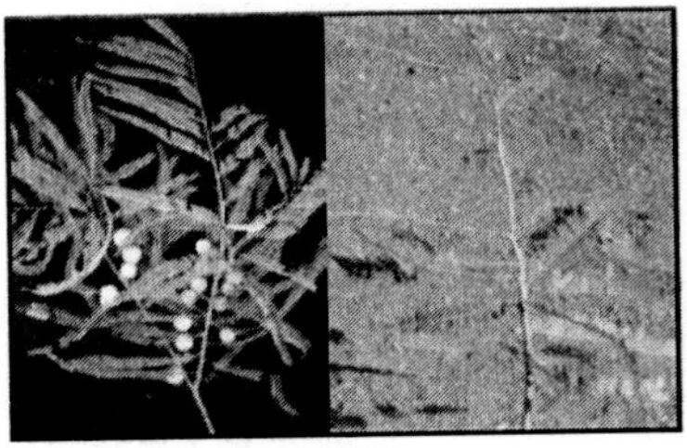

Aonla (*Emblica officinalis*)

Sounf (*Foeniculum vulgare*)

Kalihari (*Gloriosa superba*)

Ban Nimbu (*Glycosmus pentaphylla*)

Mulethi (*Glycyrriza glabra*)

Gurmar (*Gymnema sylrestre*)

Hurhur (*Gynandropsis gynandra*)

Khurasani ajwain
(*Hyoscyamus niger*)

Ratanjot (*Jatropha curcus*)

Hursagar
(*Kalancvhoe pinnata*)

Chandrasur
(*Lepidium sativum*)

German chameli
(*Matricaria chamomella*)

Lajwanti
(*Mimosa pudica*)

Chandrajot (*Jatropha podgarica*)

(*Marinus afficinalis*)

Japani pudina (*Mentha arvensis*)

Mentha

Nishod (*Operculina turpenthum*)

Flower (*Opium poppy*)

Fruit (*Opium poppy*)

Drum stick (*Moringa oleifera*)

Tulsi (*Ocimum basilicum*)

Tulsi (*Ocimum spp.*)

Shyama Tulsi (*Ocimum spp*)

Pippali (*Piper longum*)

Bhui Aonla (*Phyllanthus niruri*)

Ketki (*Pandanus odoratissimus*)

Isabgol (*Plantago ovata*)

Rajnigandha
(*Polianthus tuberosa*)

Pachouli (*Pogostemon patchouli*)

Chitrak (*Plumbago zelanika*)

Champa (*Plumeria alba*)

Bawchi
(*Psoralea corylifolia*)

(*Salvia officinalis*)

Bijasar (*Pterocarpus marsupium*)

Rose (*Rosa damascena*)

Sarpgandha (*Rauwolfia serpentina*)

Sita Ashok (*Saraka asoka*)

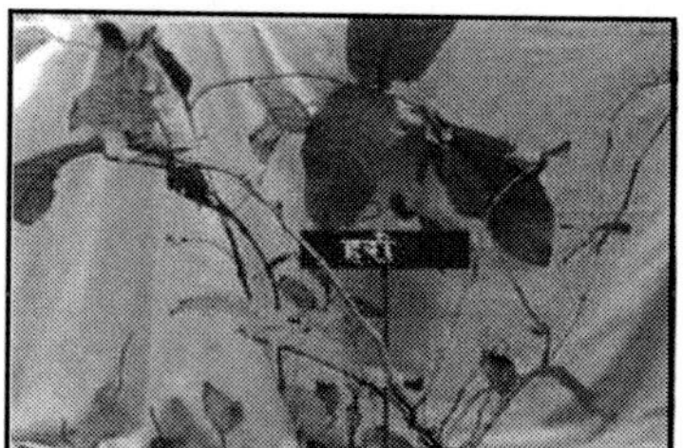

Harra (*Terminalia chebula*)

Reetha
(*Sapindus mukorossi*)

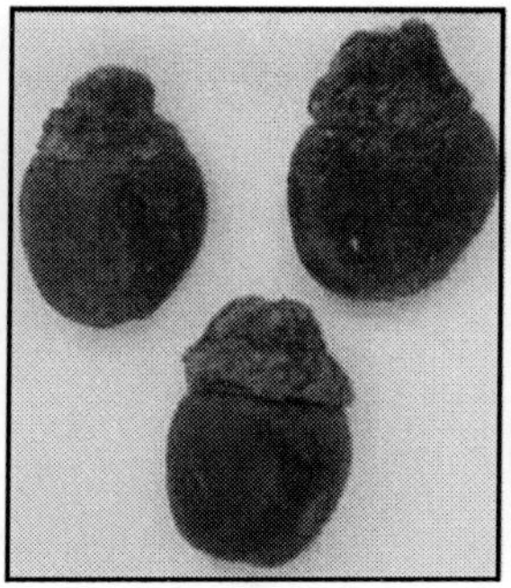

Bhilawa seed
(*Semecarpus anacardimum*)

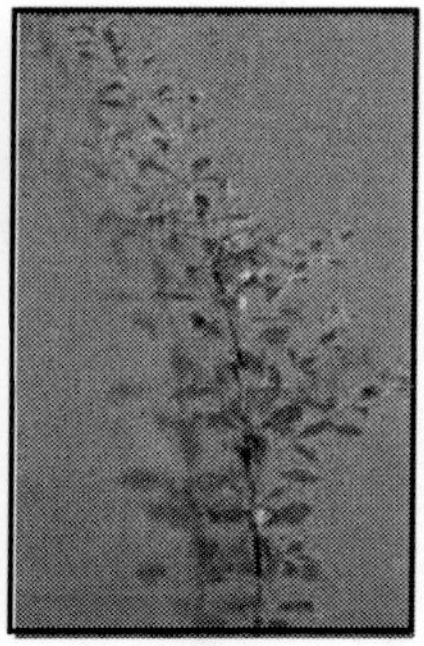

Meethi booti
(*Scorparia dulsis*)

Jangli Bhata
(*Solanum sissymbri*)

Kantkari
(*Solanum lacenisa*)

Bada Chandrika
(*Rauwolfia tetraphylla*)

Akarkara
(*Spilanthes oleracea*)

Deshi Akarkara
(*Spilanthes spp.*)

Chirayta
(*Swertia chirata*)

Marigold (*Tagetes patula*)

Bahera (*Terminalia belerica*)

Arjun (*Terminalia arjuna*)

Giloy (*Tinospora cordifolia*)

Ajmod (*Trachyspermum roxburghianum*)

Patharchata (*Trianthema monogyna*)

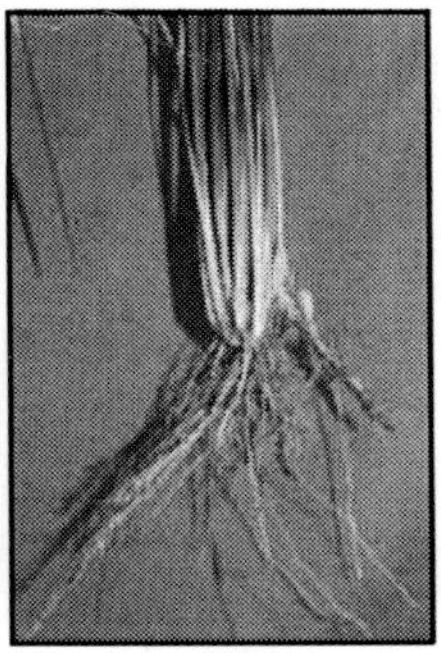

Khus (*Vetivera zizinoides*)

Ashwagandha (*Withania somnifera*)

Peela Bhrigraj
(*Wedella chinensis*)

Adrak
(*Zinziber officinale*)

Oil Distillation unit

X Governor of M.P.
Bhai Mahaveer

X Vice Chancellor JNKVV
Dr. G.B. Singh

Professor A.B. Tiwari
Explaining to farmers

Field Visit of Farmers

Dr. J. P. Tiwari
Explaining to students

Visit of Scientist to
Herbal Garden

Farmers visit to Herbal Garden

Raw Material of Medicinal Plants

Nursery Area